Japanese Corporate Transition in Time and Space

Japanese Corporate Transition in Time and Space

Tomoko Kurihara

JAPANESE CORPORATE TRANSITION IN TIME AND SPACE
Copyright © Tomoko Kurihara, 2009.

Nicolas Roope, Antirom, *RGB Show*, OneDotZero Festival, ICA, London, May 1999.

"Seken," chapter 8, reprinted by kind permission of Blackwell Publishing from *The Blackwell Encyclopedia of Sociology*, ed. George Ritzer, vol. VIII, pp. 4154–7. © Blackwell Publishing Ltd 2007.

First published in 2009 by
PALGRAVE MACMILLAN®
in the United States—a division of St. Martin's Press LLC,
175 Fifth Avenue, New York, NY 10010.

Where this book is distributed in the UK, Europe and the rest of the world, this is by Palgrave Macmillan, a division of Macmillan Publishers Limited, registered in England, company number 785998, of Houndmills, Basingstoke, Hampshire RG21 6XS.

Palgrave Macmillan is the global academic imprint of the above companies and has companies and representatives throughout the world.

Palgrave® and Macmillan® are registered trademarks in the United States, the United Kingdom, Europe and other countries.

ISBN: 978–1–4039–6654–4

Library of Congress Cataloging-in-Publication Data

Kurihara, Tomoko.
    Japanese corporate transition in time and space / Tomoko Kurihara.
        p. cm.
    Includes bibliographical references and index.
    ISBN-13: 978–1–4039–6654–4 (alk. paper)
    ISBN-10: 1–4039–6654–0 (alk. paper)
        1. Corporate culture—Japan. 2. Organizational behavior—Japan.
    I. Title.

HD58.7.K88 2009
302.3'50952—dc22   2008055686

A catalogue record of the book is available from the British Library.

Design by Newgen Imaging Systems (P) Ltd., Chennai, India.

First edition: September 2009

10 9 8 7 6 5 4 3 2 1

Printed in the United States of America.

For Emiko and Nobuhiro Kurihara,

Kurihara Nobuko and Etsuhiro,

Yuasa Kimiko and Kunio in memoria,

and Névine Z. Davies

You see, but you do not observe. The distinction is clear.

A. Conan Doyle

Beauty is truth, truth beauty,—that is all

J. Keats

Contents

Plates and Tables

Plates

Tables

Note on Transliteration and Translation

Isolated words and phrases in Japanese appear romanized following the revised Hepburn system. In addition, in relation to single words and short phrases, the Japanese script in *kanji* and *hiragana* are given in the main text if presence of the characters significantly enhances the understanding of the argument. Longer phrases, usually quotations from interviews, conversations, or text, are translated in English followed by the original in Japanese or vice versa. In this instance, no transliteration is provided. The Japanese has been retained as it is essential to convey the linguistic knowledge that forms the core of this book. All translations are mine unless otherwise stated.

Acknowledgments

This book could not have been completed without the generosity of many individuals. I was touched by and remain grateful to everyone I have encountered on this journey. Foremost, I wish to thank the managers of the personnel and overseas departments and board of directors at JCars who allowed me access to study their organization for my PhD dissertation. My internship entailed a monthly salary of ¥80,000, and daily travel expenses, and further costs incurred upon visiting other fieldwork related sites were supplemented. I should note that my writing is not in any way constrained or shaped by this financial support (all interns at JCars are salaried during placements). I thank all the employees at JCars who interrupted their work to respond generously and in depth to my repetitive and persistent inquiries; those who offered their friendship to me; and for sharing with me their environment wherein I felt free to participate. I regret that maintaining their anonymity prevents me from naming them personally. The use of pseudonyms is an ethical requirement to the practice of ethnography and in my writing I have protected the identity of JCars and the identity of individual workers. For this reason I have chosen not to include photographs in this book.

I thank Jonathan Watts, for arranging my fieldwork at his Tokyo office. I also thank his colleagues who endured my inquisitiveness during this time. I thank Yoshida Tomoko at Keiō University who kindly discussed returnee issues with me. I am also grateful to the returnees who via email and interviews took part in my study during the initial months of my fieldwork.

I wish to thank members of my family for their ongoing care, advice, and encouragement while I pursue my projects: mum and dad, Emiko and Nobuhiro Kurihara. My love of literature, autumnal leaves, and observing people, in fact, everything I think of as uniquely mine comes from you. Thank you for being my wonderful parents. My sister Utako Oyamada, and Bess-chan for keeping me in good humor. My sweet grandparents, Kurihara Nobuko and Etsuhiro. Inma Hiroko, Fujio, and family. The Motoike family. I am far away but my heart is close.

I thank all my friends for their splendid company and inspiration. In particular, I wish to thank Celayne Heaton-Shrestha for commenting on chapters 1 and 2; Vesna Antwan, Eddie Lindisfarne, Rui Da Silva, and Arjun Parekh in Japan; Nick Ryan for help with Plate designs; Nicolas Roope for help with reworking Plates and shooting the cover photograph; the wise and graceful counsel of Névine Davies, an additional thank you for commenting on chapters 2 and 5.

Writing was supported by the Economic and Social Research Council Postdoctoral Fellowship award (# T026271269). I thank Don Slater, Sociology, London School of Economics for hosting and mentoring me during this fellowship.

My four-year divagation from this book during my postdoctoral research posts at Cambridge University proved informative to my thinking: my admiration goes to Mia Gray, geography department, for her brilliance, generosity, and friendship (EU-funded project under the Information Society Technology Programme [RISESI IST-2001–33189]). I thank Brendan Burchell and David Good in the Faculty of Social and Political Sciences (study of distributed organizations funded by Cambridge-MIT Institute). It was a pleasure working with you all. I also wish to thank John Swenson-Wright, Faculty of Oriental Studies, for his exacting criticism on a draft of my manuscript.

At the School of Oriental and African Studies, London University, I wish to thank my remarkable thesis advisors Nancy Lindisfarne, Lola Martinez, and Kit Davis for their advice and guidance. I thank Harumi Befu at Kyoto Bunkyō University for his advice during my fieldwork. I thank Louella Matsunaga at Oxford Brookes University and Danny Miller at University College, London for their advice to take my thesis to publication. I am grateful to the two anonymous referees who provided useful comments on my original manuscript. Finally, I thank my editors Toby Wahl and Asa Johnson at Palgrave Macmillan for their support.

CHAPTER 1

Knowledge and Competing Discourses in Organizations

JCars in Context

JCars. A transnational corporation in Osaka, Japan. The corporate Public Relations brochure is 34 pages, printed on high-quality, A4-sized paper. It is at once a product and mode of social interaction.[1] The front cover is dusted with metallic effect. Underneath swirls a gradation of blue hues from intense to light aquamarine. Photographic images take center-stage while blocks of concise text are interspersed on every other page. Together they outline JCar's corporate history, philosophy, organization, strategy, and future orientation in relation to the global marketplace.

Mankind and Nature: A steel-grey sky meets the grey-blue wintry ocean over which wayward seagulls take flight; a germinating sapling growing amidst ancient conifers in a dense forest; schools of silver-green tropical fish diving in the ocean depths; a herd of zebra sprinting powerfully and gracefully against a blazing orange sunset. Various positive characteristics about the workplace, of nurturance, endurance, speed, energy, robustness, and freedom are actively conveyed through use of nature-based metaphors. A poetic passage draws parallels between mankind, corporation, and nature:

たくましく、豊かな恵みを与えてくれる自然。すべての生き物を慈しみ、ある時はやさしく、またある時は過酷なまでに厳しい表情を見せる自然。ひとは、そんな自然の大きな力と対峙しながら、そこに自らの夢と理想を重ね、自らが信じるものを**創り**出してきました。
　　私たちの**製品づくり**も同じです。(My emphasis)

[Nature is robust, bountiful, and provides abundantly for mankind. Nature has many faces, showing tenderness to all living beings, at times gentle, at other times ruthless in its cruelty. Mankind confronts this enormous force of nature, and by entrusting our dreams and ideals to the task, through our own accord, we have created material objects which hitherto have not existed, in these we place firm belief and confidence.

Our product manufacturing follows this philosophy.]

Global Presence: Numerous aerial photographs catalogue the company's global subsidiaries in Asia, Europe, North and South America, manufacturing and assembly plants, vast storage, and shipment areas. These images attest to the sheer scale of operations. Oddly the head office appears modest by comparison.

Advanced High-Tech Knowledge: Images relay a comprehensive range of JCars' products. The most impressive among the engineering-related photographs, even to the non-technically minded, are those that display the complex, large-scale, high-tech equipment used in Research and Development. Skill and technical knowledge are clearly central to the success of the company.

Workplace Social Relations: Meeting scenes feature an international cast engaging in discussion, relaxed and smiling. Fresh-faced youthful engineers, both female and male, wearing white lab coats cooperate closely in a friendly manner. They work unsupervised. Concentration, enthusiasm, and dynamism come across strongly. A young woman in office uniform listens to comments given by an older male colleague as he refers to a document in hand, she stands back from him and leans in slightly, he is seated. Status distinction by gender, age, and occupation are embodied very clearly here, through posture, interaction, distance, gaze, and dress.

This brochure represents the management's total vision of the company.[2] Both the brochure and vision are produced by the personnel department. The head of department, a board member, has considerable influence on corporate strategy. Sato-*torishimari-yaku* selected these nature metaphors (*inyu*) to portray a clean and likeable image of JCars. In these images, we also gain a sense of JCars' respectful and active concern for worker safety and pollution reduction in the engineering and manufacturing process. The environment and *man*ufacturing are one and this corporation's deference toward its workers and the environment conveys a humble attitude. In fact, the quoted passage conveys far more. Let us note the repeated reference to creating and manufacturing (emphasized in passage) that appears to point out more than simply what the company does; after all, this could be indicated by other Japanese characters for "create" (*tsukuru*), such as 作る and 造る. Rather, the character for *tsukuru* (創る) used in this context contains the added nuance of "being the first to create something that hitherto does not exist": effectively, "invention": this differs from, and perhaps even surpasses, the meaning of "innovation" that seeks to make changes in something already existing, by new methods and ideas. 創る: this *kanji* is evocative, particularly in the context of the passage about man and nature, and in relation to the whole brochure; it speaks of a passionate and caring attitude of the company. Furthermore, in this representation we find parallels with the ethic of craft, crafting, or craftsmanship. Effectively, the brochure outlines a set of ethical and aesthetic principles.

The anthropologist Dorrine Kondo (1990a) writes of artisanship in her ethnography of a Japanese sweet factory as "creative self-realization through work, by polishing the self through hardship" and as an aesthetic connecting nature and the material world (Kondo 1990a: 234). For the sociologist Richard Sennett (2008), having in place institutional structures that provide an environment that enables craftsmanship to flourish is paramount to the valuing of skill; and, skill, defined widely, is increasingly necessary to post-Capitalist culture of work. Like Kondo, Sennett connotes an inclusive definition of craftsmanship: namely, it is about mastery, a human impulse to do a job well for its own sake; a combination of the procedure of completion and higher level problem solving and experimental thinking; it is these skills defined widely that form an important part of learning and self-development. Thus, in conveying this ethos of craft, connecting finished work with the subject's dreams, ideals, and confidence, the JCars brochure is sending a timeless message that might speak to many workers across different societies.

Moreover, the brochure defines a cultural knowledge of the orientation of the self/JCars to the world. We might say that this cultural knowledge is rooted in a distinctive Japanese intellectual tradition, one described by Ikujiro Nonaka and Hirotaka Takeuchi (1995: 25–32)[3] as a melange of Buddhism, Confucianism, and, Western phenomenology, analytical philosophy, and pragmatism; it consists of three features centered on the oneness of (1) humanity and nature; (2) body and mind; and (3) self and other. Let us take note that in this tradition the paired elements within binaries do not oppose but rather intersect and interconnect. In fact, as many deconstructivist and practice theory-oriented theorists contend, this is the stance anthropology should take in relation to its subject (Moore and Sanders 2006).

My purpose of introducing JCars (my field site) via their PR brochure is twofold: first, to immediately immerse the reader in a tangible, rather than abstract, description of this Japanese company; second, to remind us of representation and its effects, as this is integral to the direction of my analysis, which I outline in this chapter. Representations encourage us to make visible the constructions that constitute phenomena. Simultaneously, they also remind us to cast our eye further to find the contradictions and concealments, or prohibitions of what one must not represent (Derrida 1990: 135; see also Heidegger 1962).

In this book we enter the world of JCars. This "world" is not self-enclosed, but intersecting with other companies and institutions at national and global scales (Ben-Ari 1994; Hamada 1992; Ong 1987; Saso 1990). Engaging the local with the global, as Mitch Sedgwick (2001: 43) has shown, "Japanese corporations are the central filter through which 'Japan' interacts with the world." Likewise, the transformation of corporate culture in Japan that we explore in this ethnographic study of JCars is embedded within a global context. The changes we will read about in this book are both particular to Japan and applicable more generally; they express features that are shaped in the postmodern, post-Capitalist era of global change. In this regard, we might observe that Japanese corporate culture does not differ significantly from other societies in the global economy.

This global terrain includes features such as "empire" as a new decentered and deterritorialized form of power (Hardt and Negri 2000); the control of globalization by elite members of the "transnational capitalist class" (workers in corporations) (Sklair 2001) living in "global cities" like New York, London, and Tokyo (Sassen 1990); flexible labor markets and organizations leading to "corrosion of character" from disintegration of qualities such as "loyalty, commitment, purpose and resolution" (Sennett 1998); social relationships as "networks" that enable individuals to shift between volatile connection, disconnection, and reconnection in a "liquid" world (Bauman 2003), as disjunctive "flows" and the new role of "imagination" in social life (Appadurai 1990; 1996), as hybrids of humans and technology (Latour 1996); and the proliferation of "non-places" that create a solitary contractuality in an individual's relation to others within that space (Augé 1995).

With regard to restructuring in global terms, we have witnessed significant shifts in three intersecting areas:[4] (1) production, work, and economic life—in particular, changing notions of "work" and "occupational status" as a result of changing labor market structures and diversification of employment practices, such as seen in the growth of subtypes of non-regular forms of employment due to the feminization of employment; (2) the nature of the state, power domains, new social movements, and socio-political identities—in particular, new social identities among the workplace; and (3) knowledge, science, and technology—how technology reshapes social relations and dynamics within the workplace community.

The JCars brochure continues to circulate within this global context. My fieldwork was conducted in 1998–9. Fortuitously it coincided with the run-up to the enforcement (April 1999) of the revised Equal Employment Opportunity Law (EEOL) (enacted in June 1997) and the installment of IT infrastructure in JCars (six months prior to my arrival as fieldworker individuals were issued with laptop computers providing continual access to email and Intranet). During 1998–9, the unemployment rate was 4.2 percent; the real economic growth rate −2 percent; and the ratio of active job openings to active job applicants 0.48, the lowest on record in Japan (Japan

Institute for Labour Policy and Training 2006: 12). Therefore, my field-work was an opportune and unusual time to capture the structural shifts of the 1990s, in labor markets and the employment situation in businesses, and, in cultural changes in wider society affecting working life. This book then examines the changes to women's status in workplaces and the transition of social and cultural values that reshape workers' awareness and needs, which in turn impact the way management maintains the workplace community.

Japan's socio-economic and cultural transition has its origins in the 1980s, which led to the collapse of the bubble economy in the early 1990s, and the long-term recession during the 1990s. The recession hit bottom in 2002, but since then the Japanese economy has recovered steadily.

During the late 1990s the economy destabilized people's lives and identities across all sections of Japanese society: we observed the structural breakdown of contemporary Japanese society in all institutions—work, home, education—and experienced this in political, economic, social, and cultural spheres (see Nathan 2004; Yoda 2006; Zielenziger 2006).[5] Notably, social affluence was accompanied by a moral decline that was observed in the weakening of postwar values and norms, and, work ethic; respect for hierarchy and authority; loose sexual mores among youths; a rise in increasingly individuated and fragmented lifestyles; increase in the numbers of homeless; and, a rise in suicide rates, particularly among men.

Increased socio-economic inequality is related to the combined effect of several factors: an aging society confronted by mass retirement of the first baby boom generation; a declining birth rate due to later marriages, growth in numbers of unmarried singles, and the phenomenon of "parasite-singles" living with parents; and increasing divorce rates.[6] Structural changes in labor markets also contribute to this widening inequality: with low hiring rates; uncertainty in regular workers' wages, benefits, and job security; and increase in non-regular workers following the liberalization of the temporary market, and subsequent diversification of employment mainly among women and young people (*neet* and *freeters*).

Notably, the prolonged recession led to the structural breakdown of one of Japan's fundamental institutions, enterprise society (*kigyō shakai*) or corporate-centered society (*kigyō-chūshin shakai*): the state-sponsored and protected, postwar, patriarchal project of modernization, made up of a vast network of companies headed by large firms in the private sector in the manufacturing industries. The individual's loss of freedom due to corporations taking over workers' lives characterizes this society. Patriarchy is another characteristic; the government designed the sustainability of corporate-centered society on social security policies that enforce a gendered division of labor and segregation of labor markets by gender (Osawa Mari 1994).[7]

While the following chapters investigate how these social and cultural changes impact JCars, in the remainder of this chapter, I discuss how my approach relates to previous works by anthropologists of Japanese workplaces.

Models of the Japanese Workplace

The workplace is represented by (1) anthropologists and sociologists, who draw on perspectives of (2) the management, through its ideals and ideologies that are necessary to lead the organization; and (3) lived experience of workers and management. We can say that (1) and (2) overlap: represented in views of structure, rules, and ideology; and (2) and (3) overlap in a state of discursive construction, represented by (1) as the space of structure and practice, as reality, as contestation, as negotiable, in the form of constraints and agency. Contemporary ethnographies since the 1980s and 1990s critically examine levels (2) and (3) and their meaning in fine detail, in the contexts in which they arise. However, the matter of how we balance the perspectives of (2) and (3) in our accounts still demands attention. We think through this problem in this section.

Two important frameworks have been central to our understanding of Japanese workplaces: The first, known as the group model, emphasizes the vertical structure of society (Abegglen 1958; Clark 1979; Dore 1987; Johnson 1982; Nakane 1970; Reischauer 1977). Prominence is accorded to the role of hierarchical relationships in shaping interpersonal relations and the relations between departmental sections within the organization. Notions of harmony and consensus inform the qualitative descriptions of social relations outlined in this model.

However, vertical and horizontal relationships in practice actually carry numerous nuances, beyond those that the group model is able to acknowledge (Atsumi 1980). In other words, there are multiple ways of indexing the distance between individuals situated in these rigid, seemingly rule-bound relationships (Bachnik and Quinn eds. 1994; Bachnik 1998: 111).

The second is a riposte to the group model. This model attributes agency to individuals when accounting for social action and takes account of symbolic action. There are two orientations: the first acknowledges the existence of conflict (Befu 1980, 1989; Eisenstadt and Ben-Ari eds. 1990; Krauss et al. eds. 1984; Mouer and Sugimoto 1980; Mouer and Sugimoto eds. 1989; Pharr 1990); the second looks to acts of resistance (Lo 1990; Ogasawara 1998; Saso 1990).

Both the conflict and resistance approaches seek to contest the dominance of structure. Paradoxically, both disappoint us because, conflict and resistance describe social action based on interpreting action against social structure and this positions subjects vis-à-vis authority in a way that ensures the reassertion of the ideology of hierarchical power (cf., Brown 1996). Thereby, unable to break from it, structure/harmony remains their raison d'être. The group model and conflict and resistance models are thus very much linked in terms of development, where the one is rooted in the other. Models must simplify in order to gain explanatory power, and this involves eliminating other perspectives in support of the chosen view (Appadurai 1988), so it is inevitable that models essentialize to some degree. The focus on conflict and hegemony dispenses with other aspects of social life such as cooperation and reciprocity, although these characteristics together form the nature of social life (Ortner 1984).

Dorrine Kondo (1990a) argues that resistance fails to account for the complexities of human agency, the "*multiplicity* of sometimes paradoxical and creative effects" produced under power (p. 225, italics in original). Kondo advocates capturing meaning not through foundational categories of resistance and power but rather through our subjects' discursive strategies (ibid.). The discursive approach effectively opens up potentials for cultural and social transformation, as forces of opposition exist in relation to the other within a processual dynamic.

All writing is *intertextual* (Kristeva 1980: 66). Thus, building on the insights into Japanese work life described and analyzed in the ethnographic work that precedes mine, I too begin from a place that recognizes coexistence:

> What is so often missed in discussion of community in Japan, especially workplace community, is the coexistence of dissonance and harmony, of often frustrating relationships of unequal power and other relationships of equality, and of feelings of being isolated or taken advantage of with feelings of comfort and caring. (Turner 1995: 94)

While recognizing the coexistence of dissonance and harmony, my text begins from a supposition that the workplace community is founded on discordant relations, where "instability, uncertainty, and dissensus" characterize organizations in which "differentiation" is a key feature of social relations (Grint 1991: 145). Differentiation also characterizes the wider society; Japan is described as an increasingly fragmented society and differentiation can be observed, for instance, in the transformation of labor market structure. From disorderly (*midare* or *midareru*) relations in the workplace, the management, and workers then attempt to shape order, harmony, and their ideal community. Hierarchy is thus central to the management's mission.

A further point in relation to coexistence concerns my attempt to link theory with my subject's practices observed in the field; this involves the resolution of the conventional oppositions that structure culturally dominant representations with regard to gender, identity, status, and work.

These are men versus women; career track versus clerical track workers; inside versus outside; management versus employees; work versus play and other non-work activities; and, formal versus informal spheres of activity. Putting opposites in dialogue with each other does not imply their disappearance but rather underscores coexistence (see Allison 1994; Bachnik 1998; Turner 1995). Our experience of the social world is not confined to one position at either end of the binary: this position is imposed on subjects through reification of categories as objects of representation and analysis. Therefore, my task is to assess how knowledge is produced through the dialectical tension between the coexisting categories.

Below I propose how to study the workplace community through discursive practices.

The Workplace as Community

JCars, like any large company, can be studied as a firm community (*kaisha kyōdōtai*). This term implies a group with its unique set of boundaries defining social norms and moral constraints that act on its employees (Suehiro 1998). The firm community is thought to structure corporate-centered society and vice versa (Tabata 1998); in this way the two terms are related, but they are not the same.

From an anthropological and sociological perspective, the workplace can be described as a community because it is an "organized" institution, characterized by an internal structure that organizes social relationships within the workplace (Warner 1941), unified by a "community of interests" (Williams 1988: 75) that has "defined limits to membership" (Layton 1997: 39). This traditional understanding of the term is useful in framing the workplace as a community, but what we wish to take note of concerns the degree of unity observed, how "interests" are defined, by whom, and for what purpose.

Tönnies (1955) outlines the qualitative characteristics of community (*gemeinschaft*), as dominated by kinship and moral bonds, giving rise to a homogeneous, cohesive, and traditional social order. Communities have therefore tended to be seen as social facts, as things-in-themselves, *á la* Durkheim (Durkheim and Mauss 1969). Tönnies' description of community is comparable to the way Japanese companies have been described, often through the metaphor of kinship, specifically as ritual-kinship, as constituting a family or "*ie*" (Bennett and Ishino 1963; Kondo 1990a; Nakane 1970; Noguchi 1990; Vogel 1963).

The idea of workplace as family is drawn heavily from the management's perspective of the ideal workplace, rather than from the workers' views; to this extent, it is an ideal that can be contested (Kondo 1990a; Noguchi 1990). The *ie* metaphor is still in use in an ironic sense (Kondo

1990a). This shows how the company as family system, which engenders qualities of harmony and cooperation, has relinquished some of its credibility through the recognition that an all-encompassing definition glosses over the quality or type of social relationships found in the workplace. Yet, familial metaphors are still in use. For example, relations between female colleagues at JCars unfold as a sibling relationship of "sisters," which defines the vertical relationship between seniors and juniors with respect to tutorage.

The strength of interpersonal bonds within the workplace community is a basic assumption of Japanese companies, for example, in comparisons between Japanese, American, and British workplaces, Japanese workers and management are more often said to share in a "strong sense of participation in the company as a 'community of fate'" (Cole 1971: 240; cf., Nakane 1970); where the "... member ... disregards his independence ... and accepts the burdens of responsibility as a participant" (Rohlen 1974). However, if devotion is rooted in economic motivation of individuals, then, "common interests" are not commensurate with collectivist emotion (Tabata 1998). Thus we might share in Parson's idea that solidarity within groups is "... a function of the level and distribution of loyalty of its members but not synonymous with it" (Parsons 1975: 60). Overemphasizing the strength of interpersonal bonds within the workplace community is also problematic for the same reason as the application of the family metaphor: it is a view promoted largely by management. Moreover, there are varying degrees of commitment and participation to the workplace; this shifts, being more pronounced in times of social and cultural transition (Hamada 1996).

I prefer to follow a notion of community that is not based on predetermined notions of strong interpersonal ties and solidarity uniting fellow workers and management, but on a loose notion of an existing feeling of communion, which is "imagined" and "created" (Anderson 1991) through discursive practices. Benedict Anderson's idea of "imagined communities" enables a discussion of the "style" in which the community is conceptualized and represented. Furthermore, a symbolic analysis of a community that pays close attention to the negotiation and contestation of meaning (Cohen 1985) is vital to this study. How, then, can this question of *style* be analyzed?

Discursive Formation of Community

What is the value of the concept of "discourse" for detecting the style in which the community is conceptualized and represented? Through discourse, I believe we gain an understanding of what is at stake (knowledge, power, status, ideology, or identity) at different levels (gendered, hierarchical, between peers). Moreover, we restore dignity to the subject who is caught up in the web of knowledge and power (see chapter 4, n. 20). In my analysis, I invoke concepts from a range of theorists; my eclectic approach reflects how theorists have used discourse to mean different things, which in turn shows the unique value of each concept for explaining phenomena at different levels. I touch lightly on the theory here as the utility of theory is best shown in exchange with field data. I use the terms "management discourse," "discourse of workers," "practice of workers" and "practice of management," interchangeably, to indicate an established way of talking about and actualizing some idea, in this case, the shaping of the workplace as a community.

The workplace community is conceptualized in various ways. It always involves knowledge and power relations; these are articulated through the way managerial logic shapes notions of the ideal workplace as well as through the subjects' lens, in how the notion of community is perceived and spoken about. A discursive approach enables an account of these two levels, analyzing the ways in which they interact, reinforce, and conflict with the other, because ". . . discourses . . . [are] practices that systematically form the objects of which they speak" (Foucault 1972: 49). Michel Foucault's notion of discourse is useful, in the way it brings together statements made by various individuals, while enabling access to a particular discursive object through the link formed between the statement and its subjects, objects, and concepts (Deleuze 1988: 4–10).

Foucault's notion of discourse, then, as a structured framework through which workers perceive the workplace is useful, as on one level, it enables reference to the techniques used by management to shape the workplace, a force akin to ideology (Foucault 1972: 38), without the class-based, top-down, oppressive connotations associated with the traditional Marxist understanding of ideology. The meaning of ideology is wide-ranging and refers to a "collective symbolic production" of "an articulated system of meanings" that is "the possession of one social group" (Comaroff and Comaroff 1991); it can also mean "the establishment and defence of patterns of belief and value" that might be shared by both the superordinate and subordinate group (Geertz 1973a: 231).

By ideology I am not referring to political action at a macro level based on class-based assumptions, because postwar Japanese history has not been centered on class struggles as Europe's has; rather social stratification proceeds on the basis of status differences induced by education, gender, and age (Hamada 1985; Kondo 1990a; Sugimoto 1997). I do, however, accept that ideology refers to a cultural process necessary for defining a community. Management discourse/practice thus refers to the styles of representation observable in the process of signification of ideas, beliefs, and values articulated in various contexts within social life (see Eagleton 1991: 1–31). In connection, to describe the more explicit, yet, symbolic way the management discourse/practice/power functions in the workplace, particularly from the management side, I borrow the notion of *government* (Foucault 1997) that means to "lead," which structures, forms, shapes, or guides the "possible field of action of others" (p. 26).

The actual processes by which management inculcates ideas among workers is best described through Pierre Bourdieu's notion of *habitus* (1977; 1990) that accounts for the tacit experience of structure through the overall framework of practice. Although *habitus* is experienced unconsciously, the social world can only be experienced through the body (Bourdieu 1984). Moreover, as experience is a conscious process, we might say that *habitus* constitutes an experiential notion. Thus, *habitus* describes an embodied way of knowing the world, of knowing social relationships; it bridges rules and reasons given to explain social action (Taylor 1999). I will use *habitus* to account for the managers and workers perspectives.[8]

Importantly, management discourse/practice in relation to implied politics or authority emphasizes a mutual process, whereby both managers and workers are caught up in micro political processes in the office. To describe the multiplicity of subject positions present within discourse, particularly with respect to status negotiation, I draw on Mikhail Bakhtin's (1981) concept of *heteroglossia*. Bakhtin's liberationalist, optimistic, processual approach allows for the inclusion of the marginal, decentered reading of power. His view of discourse is inclusive by its totality, diversity of ideas, and voices captured in it:

> ... [D]iversity of social speech types (sometimes even diversity of languages) and a diversity of individual voices.... The internal stratification of any single national language into social dialects, characteristic group behaviour, professional jargons, generic languages, languages of generations and age groups, tendentious languages, languages of the authorities, of various circles and of passing fashions, languages that serve the specific sociopolitical purposes of the day, even of the hour (each day has its own slogan, its own vocabulary, its own emphases)—this internal stratification present in any language at any given moment of its historical existence ... (pp. 262–3)

Bakhtin's emphasis on the every day is similar to that of Michel de Certeau (1984), whose theory of tactics, "making do" and the efforts to repress or conceal it (*la perruque*) is also useful to our theorizing. De Certeau's approach supplements Bourdieu's: it frees us from Bourdieu's subtle structuralism; and from the abstraction and fluency of *habitus* in shaping social action, which makes no account of "microprocesses of cognition in the world of practice" (Margolis 1999: 78).

We also wish to recognize the materiality of discourse in that discourse not only refers to language or speech.[9] Indeed, practice incorporates discourses (de Certeau 1984), in other words, utterances are performative in that saying is a form of doing (Austin 1962). Thus, it is useful to think of language as presenting the body in its action (Butler 2004); after all, speech is a social action that generates effects, without which social transformation cannot occur (Butler 1999).

In summary, important to my approach is the essential understanding that every subject is a producer of knowledge—a knowledge of company ideology; of each other; gendered relations; temporalities and spatial orientation—and their voices overlap or exist in tension with one another in an ongoing process of negotiation within discursive practices. I strive to be attentive to every (different) voice present in the workplace—different levels of discourse and multiple perspectives—and integrate them in my text. Respecting the contexts from which they emerge is equally critical. From *where* the subject speaks is more, or as, important as what is said. In effect, two kinds of knowledge inform this study: knowledge expressed in language and embodied knowledge. In writing about white-collar workers, I avoid taking any aspect of their every day for granted, and hope to convey what is vital to their lives.

Hierarchy in Action

In exploring the various symbolic means through which knowledge of community is enacted, we revisit hierarchy because hierarchy is in fact central to the management's mission of shaping the ideal community. Thus, we look at "hierarchy in action" (Smith 1983: 48) through the daily lives of workers (Rohlen 1974) thereby drawing on the creative and flexible aspects of Japanese organizations. In fact, it is useful to problematize the very concept of organization: "organization as culture" as opposed to "the culture of the organization": it alerts us to an important point, that meaning making is a continual process (Kondo 1990a) and many cultures exist within any one organization (Wright 1994: 19). Following a processual, practice-oriented approach, I analyze the notion of hierarchy in the context of the workplace, looking at the ways through which it is constituted and reconstituted in social practice. We will recognize, too, that the dominant discourse or representations presented by the company to the outside and to its workers mask the internal and generic contradictions and inconsistencies of its own discourse.

With a deconstructive and discursive framework firmly in mind, I am most concerned to destabilize the hegemony of cultural ideals and corporate ideology by unravelling the structures and practices that alternately endorse and subvert the appearance and maintenance of a coherent, seemingly natural state, that is, an ideal workforce and workplace community. Hierarchy (by gender, tenure, and age) is instituted as a structure or principle that organizes social relationships, but it is activated differently in practice, and this leads to unforeseen and unpredictable outcomes that then have to be managed. All workers, not only managers, have to manage social relations and status positions. Hierarchy uses ideals, differences, and oppositions to shape working time and spaces, but it also mutates and becomes inconsistent with the ideal that it is intended to represent. These broad themes centered on the experience of hierarchy explored at a time of social and cultural change form the unique core of this book.

In summary, my aims are to bring a deconstructive yet processual approach into conversation with the anthropology of Japanese organizations; link structure and practice, and to reconcile theory with practice in ethnography; expose the variability in knowledge of my subjects that produce varied experiences and identity, while also noting the normative, expected conduct that orients them; show how the category of gender intersects with other relevant categories such as age, class, status, occupational role, and appearance; restore binary oppositions that structure analyses in the hope of opening up possibilities for new understanding; and, analyze gender and authority (embodied in hierarchy), and time and space, in its performative and embodied dimensions.

Organization of This Book

This book provides an examination of new spaces and topics from which to understand corporate life; these are

- empirical data on the efficacy of the revised 1997 EEOL via the analysis of employment practices (hiring, changes to occupational role, promotion, career pathways);
- analysis of the impact of legal and social changes on the status of women;
- examination of the place of disabled workers in an elite corporation, and the wider employment practices with respect to disability in Japanese businesses;
- analysis of the complex landscape of emotions and linguistic practices in a (disorderly) community;
- examination of how the management constructs an image of community and enforces ideals through symbolic means and repeated acts such as the commute, overtime, and gift-giving;
- past ethnographies have been shy of analyzing the spatial dimensions of office life. I demonstrate that spatial practices in workplaces offer analytical purchase for an examination of the operation of hierarchy and status negotiation;
- ethnographies of Internet use in Japanese white-collar organizations are not commonplace. This book makes a significant contribution to an

emerging field that studies new technologies in the context of local knowledge;
- analysis of friendships and romantic relationships in a Japanese white-collar office.

This book is organized so that each chapter tells its own story by engaging with a thematic topic that coincides with observations in the field. It follows that the theoretical and ethnographic literature is reviewed in each chapter.

In chapter 2, I outline my fieldwork at JCars, describing my daily commute; this leads to a description of the office. I explain how I came to be a worker at this office and outline the particular circumstances that defined my situatedness. Data gathering techniques (participant observation and interviewing) in corporate research are discussed in detail.

Chapter 3 provides an overview of macro perspectives, such as the origins of gender segregation in labor markets, social and organizational norms concerning gender, economic conditions, and the judicial requirements of the 1997 Equal Employment Opportunity Law that propose equality of gender relations within organizations. The status of women is the underlying theme.

In chapter 4, we explore changes to the employment system in the 1990s via how social relations at JCars are structured through the recruitment process and tracking system (allocation and conversion), and more broadly through charting career paths. In this context, we discuss the manifold ways by which status is negotiated: through work content or occupational roles; femininity and masculinity; physical appearance; personality; and, disability. We then consider the changing image of office ladies as a social category.

Chapter 5 delves into a detailed exploration of the complex relations between senior (*senpai*) and junior (*kōhai*) workers, and between individuals of the same stratum (*dōki* group). Nakane Chie's (1970) model of social structure is discussed owing to the fact that her model is reflected in the management's ideals concerning the quality of interrelationships between workers. We also examine the experience of sisterhood in vertical relationships between female co-workers. Furthermore, we will be interested in how these interrelationships are lived and expressed through language.

Expanding on our analysis of whether the property of hierarchy itself is synonymous with harmony, in chapter 6, we analyze the symbolic practices that constitute the workplace as a community. We see how the symbolic community is bound by shared experiences, expressed in relation to time, practices such as gift-giving and departmental events. We explore how this management discourse, which guides the evocation of a sense of community, can be reproduced as well as contested in the daily practices of workers.

In chapter 7, we examine spatial practices and the political dimensions of hierarchy. Here, previously discussed themes are brought together: chapters 5 and 6 (the critique of models of hierarchy and the articulation of ideas about community) and chapters 3 and 4 (issues of men and women's status and careers). We consider the characteristics of hierarchy when situated in the space of the office: is it rigid, or is it fluid? When and how is hierarchy rigid/fluid? We experience the operation of hierarchy in space via two intertwining perspectives of management and workers.

In chapter 8 we turn our attention to additional spatial practices: the use of email and the role of bulletin boards—these are performative spaces that supplement the normative space of the office. We ask the following questions: how does management evaluate this space and its potential effects on its position of authority? How do workers think of this space and how and for what purpose do they make use of it? How is hierarchy transformed in this alternative social space? Does hierarchy here assume the same role in surveillance as it does in the open plan office? How dis/similar are social relations in this space created by email compared to that observed in face-to-face encounters? We then contemplate office romance and friendship, the ideal worker, the ideals of community. By the end of this chapter, we begin to tie up the points made in this ethnography.

The conclusion recounts the main ethnographic points made and I provide a supplement to the field data from a visit to JCars in 2003.

CHAPTER 2

Fieldwork and Methodology

Bodies and Knowledge

JCars is a large Japanese multinational manufacturer (with around 30,000 staff). The head office is based in the bustling northern quarter of central Osaka City with the site limited to clerical work. During fieldwork,[1] living alone in my parents' home in Nara City (Nara Prefecture), which lies in a valley to the Southeast of Osaka City, for 12 months, from February 1998 to February 1999 (with a month's break in May), following a precise daily routine, I commuted into Osaka City every weekday as a full-time office worker (9AM to 6PM). I invite you, dear reader, to accompany me on this journey. The routine enforced on our bodies is important insofar as it gets us to work on time, and that it puts us in a certain productive and motivated frame of mind, but contained in such commonplace are thoughts, feelings, and activities that escape its grip; we are never simply at the mercy of spatiotemporal coercion endemic to corporate life. My experience and skills of the commute (learned semiconsciously, with an awareness of cultural differences that structure this learning) can be read to mirror that of other members of JCars as well as the commuter body as a whole. I do not claim to know other commuters' thoughts or that my thoughts would resemble theirs accurately; nonetheless there is a certain commonality of bodily experience. Thus, by this description, I bring together the particular and the general, the micro and the macro, or the individual and common experience: the dual perspective that tends to orient studies of everyday phenomena (Highmore 2002: 5). Then upon our arrival at the office, I embark on a discussion of my fieldsite, my relation to my subjects, in order to draw out some methodological points. As a general note on preparing a methodology, it might be said that all ethnographies leave emotional and cognitive imprints and these precious nuggets of inspiration become tools from which we devise our own methodology in situ.

In this chapter, further to the discussion of discursive forms of knowledge described previously, I am interested to bring in the employees as well as myself, here not so much to take account of the complexity of the workers' subjectivities and experiences (Cohen 1992) as the rest of the ethnography is devoted to that aim, nor to give an existential exposé of myself, but to rather attend to the issue of embodiment and production of anthropological and socio-cultural knowledge. Our bodies as the means of communication with the world are present in all experience and thoughts (Merleau-Ponty 2005). As our body, personality, and social positioning are conduits for a project that aspires to understand the experience of others (Goffman 1989), one way of understanding another person's (cultural) experience is to understand them through our own personal experience in the field.

Intensely connected to our experience is, of course, our writing; our personal experience as anthropologists undertaking research in the field directly produces theory and analysis (Okely 1992). Such issues were taken up in the Writing Culture debate (Clifford and Marcus 1986) that paved the way for a new writing style—no longer are accounts written by the disembodied authoritative anthropologist; rather ethnography became situated, multivoiced, and reflexive. This narrative voice of the ethnographic text differentiates the anthropological project from other social sciences:

> ...Today, anthropologists strive to explain the stages through which they have arrived at what they think. They try to make explicit their shuttling back and forth between theory and "field." The work is no longer a matter of overflying the experience of the actors at high altitude, but of restoring the located and interactive character of ethnography. Their texts give more space to other voices than to the researcher's own: voices from the archives, those of interlocutors in the field, of philosophers, literary theorists, writers. Closer attention is given to social interactions, to the anthropology of speech and other modes of communication. It is accepted, in effect, that every statement relates to a context, is contingent on the personalities of the researcher and the informants, is subject to variations caused by a wide range of factors. And these days, the authors of anthropological texts are more likely than ever before to have to explain themselves; their divine right to exercise total control over their own narrative is being contested....Reflexivity—the researcher's scrutiny brought to bear on himself, the attempt to objectivize his own subjectivity—has become the main requirement of research. (Augé and Colleyn 2006: 105–6)

Our concern with writing has again moved forward, beyond the problems of textual production, to grapple with the changing conditions under which ethnographies are conducted and produced (Marcus 1999): In contemporary society multiple layers of representations—in images and writing—pre-exist regarding the ethnographic subject. These simultaneously undermine and side with the project, altering the anthropologist's position in relation to the subject. The anthropological subject (not only the anthropologist) is a reflexive subject, highly aware of what is said and written about them. For example, my male subjects in particular, in the early stage of fieldwork, regularly prefaced their statements with a defensive or apologetic commentary for, I think, no other purpose than to display their knowledge of the way Japanese society, and specifically workplace gender relations, is debated and represented in Japan and conversely by Western media. Typical introductory statements began with, "it must be strange for you, Kurihara-san, but …"; "I'm sure you don't think this way in the West, but …"; "well, Japan should be better but …" This condition gives anthropologists a different relation to knowledge.

Bodies, our own as well as our subject's, enable access to multiple knowledges, because bodies are caught up in a wide range of forces—cultural or symbolic, social, political, economic, and moral—both at local and global levels and at the interstices. The body experiences the connection between physical and social spaces and reads its position in relation to the structures of the group (Bourdieu 1990). The body therefore gives insight into the nature of social relations; between individuals, and between the individual and culture. Bodies thereby comment on social values, norms, and boundaries.

Words and idioms convey the oneness of bodies and knowledge: Bodies (身, *mi*; 体, *karada* or *tai*) and social status (身分, *mibun*); bodies and social situation (身の立場, *mi no tachiba*); and experience as necessarily embodied (体験, *taiken*). The phrase "put oneself in another's place" is expressed as "人の身になる, *hito no mi ni naru.*" Knowledge as skills and technique is acquired through the body. As Dorrine Kondo (1990a: 238) shows in her rich ethnography of artisans, skills are not only learnt cognitively but physically: "to learn with the body" (体で覚える, *karada de oboeru*) and "attaching the technique to the body" (技術を身に付ける, *gijyutsu o mi ni tsukeru*). The body enters into language in a way that it does (can)not in English, for example, apprenticeship. This is also true of linguistic expressions of knowledge of social differences that are felt through the body. In my fieldwork, Li-san "felt with her body the differences between Japanese and Chinese society" (日本と中国の違いを身で感じた, *nihon to chūgoku no chigai o mi de kanjita*). Furthermore, intense feelings, deeply felt, permeate the body: 身に染みる, *mi ni shimiru.*

Moreover, bodies and identity are mutually constituted, and incorporate categorizations such as gender, race, class, sexuality, and age. Sharon Traweek (1995) dissects the dynamics among Japanese and American physicists, who in the same laboratory experience conflict over procedures for measuring radiation exposure on the body (the frequency of testing and safe base levels). She shows us how conflicts over bodies reveal the issues at stake, namely trust, risk, and competence. They also denote any underlying cultural dualities. For example, at a time when radiation levels in the work environment were contested, by showing up at the laboratory when pregnant, Traweek is seen by American physicists to be supporting Japanese physicists and accepting their standards of infrequent and low base levels for measurement. Bodily metaphors highlight cross-cultural differences in leadership styles. Furthermore, discussion of bodies and sports as often engaged in by the American scientists facilitate the perpetual ranking of race.

Through incorporating the anthropologist's and subject's bodies into her writing, Traweek gains access to a "certain kind of privileged knowledge" that is gendered as well as cultural. Thus, Traweek urges anthropologists to present their own body in the ethnography, to appear in the narration of situations that give rise to specific experiences in order better to evaluate the situation, included in which is what each party makes of the other.

To summarize, bodies enable theorization of both socially situated subjectivity and the values that structure it, such as moral and ethical laws, aesthetic order and power relations. My account incorporates these concerns, thus describing and theorizing bodies and their interaction in relation to Japanese society and office culture—in rituals and office space; one cannot overlook bodies!

I move on now to discuss the commute, bodies in the office, my thoughts and feelings about my subjects and their views of me, our behaviors, and how I used my body to draw out what was "normal." My aims are twofold: first, to point up the methods through which social life at JCars is explored; second, by outlining how I was situated, I show how my fieldwork experience developed into my particular research themes.

Tsūkin (Commute)

Commuter bodies and uniformed school children wearily sway with wrists in rows, elegantly draped through plastic loops of *tsuri-kawa* (overhead hand straps), confined in uncomfortable proximity to adjacent bodies that occupy every space within the carriage. Amidst the timid and tired scuffle of feet shifting positions on this 40-minute journey, occasional bursts of casual and polite conversation as well as intriguing monologues of mobile phone users can be heard. Hands turn pages of print, most often the *Nikkei* and *Asahi Shinbun* or tabloids such as *Suponichi*, pocketsize novels and textbooks. The hour: 07:52AM.

The Kinki Nihon Tetsudō (Nara-Line) express commuter train departs from Nara station. It is located underground on the edge of the old city center. As the train speeds westward toward Osaka, it leaves behind the city center, cradled in a low-rising crescent of hills. Emerging on to street level, the train passes through the grounds of the ancient Heijo palace from the time when Nara was the capital of Japan. The palace itself was eaten by fire centuries ago and what remains of those times is a Titian-red wooden structure: the Suzaku-Mon (gate), 1,700 years old and standing at 20 meters. Archaeologists have reconstructed the location of the palace by erecting large concrete platforms in the long grass that grows unrestrained. Wide, tree-lined pathways laid out in a geometric pattern also cut across the grounds. Some of my childhood memories linger here, taking part in a history that demarcates a place. Then passing Saidaiji station, narrow twisted streets of old neighborhoods accommodate concrete apartments, paddy fields, and new sprawling shopping complexes. A fantastical mock mediaeval castle, a love hotel, and a giant, gloved hand and bowling pin on the same scale salute the skyline. A further 20 minutes into the journey, sizeable houses, strings of chic cafés, and tennis courts in the exclusive residential area around Ikoma station contrast with the leisure function of the Ikoma mountain footpaths and amusement park that draw many tourists to the area.

Crossing the prefectural border into Osaka, the ground rapidly drops away as the tracks remain high above the concrete vista below. The eye is overwhelmed both by the breathtaking complexity of urban landscapes and its monochrome monotony. Here the housing becomes noticeably denser where high-rise apartment blocks of colossal proportions intermingle in functional disunity with large shops, schools, and industrial buildings. This spectacular transition of landscape emancipates the body, and somewhere between objective reference and abstraction mingle two faces of urbanity, the ethereal and the horrible. The *futon* (bedding), put out to dry on apartment balconies, paste a patchwork of colors upon the grey horizon, as though to reassure the viewer of the life taking place within these compact cells. This cityscape evokes *Floretinisches Villen Viertel* (1926), Paul Klee's post-Cubist painting (Centre Georges Pompidou, Paris, France). Now approaching the traditional *shita-machi* area (the traditional shopping and entertainment district of a city, which is usually geographically low lying) of Tsuruhashi station, the tracks once again on ground level, the urban scene reverts to the old crowded neighborhoods of wooden construction. The train coasts through the underground stations of Nihonbashi, and Uehonmachi, before sliding punctually to its destination at Namba station in the heart of Osaka City. The time: 08:30AM.

Many thousands of commuters from the surrounding prefectures of Nara, Hyogo, Kyoto, and Wakayama (Kansai region) pour into Namba station. While a few older more supercilious businessmen irked by daily life and an occasional senior citizen loiter in the rows of blue plastic chairs, smoking and drinking canned coffee or energy supplement drinks, most charge toward the subway network. Passing hurriedly through the well-maintained, high-speed automated barriers that "thank you" as they churn out commuters' travel passes, with the pounding of leather soles, the blurred phalanx of navy and grey suits rush to reach the office on time.

All train and subway platforms are marked to indicate the position of the door on arrival; there is no disorderly queuing, nor room for ambiguity to doubt the order of things. Thus, people wait to board in two neat rows. In these queues I often recognize other JCar employees. It is 08:35AM and the underground trains shoot into the platforms in alarming yet gratifying succession, that is, every two minutes. Despite the commonality of commuter experience everywhere in the world, the contraries of chaotic overcrowding regulated by speed, order, and coordination does not fail to impress upon the commuter something particular about the way of life in Japan. The workers slowly emerge onto street level, they ride the escalator, coaxed and soothed by a familiar ambient female voice drifting softly from the built-in speakers. "The incarceration-vacation is over" (de Certeau 1984: 114). The office door beckons, 200 meters in the distance.

Kaisha (Company)

At the entrance to JCars, I noted the sober company logo and observed the display of product samples and photographs. I am still mesmerized and distracted by the lingering effect of the effervescent thrill experienced while walking through the underground network of trendy shops and cafés. The heavy glass doors yield willingly as I pass through; on the left is a corner for public phones, beyond which I can see three elevators. Across from these, the reception area beckons. It is set apart from this more public area by more glass doors. A young woman stands at the reception counter, her face framed nicely by her black bobbed hair with a heavy fringe, wearing a matching navy waistcoat and skirt, white blouse and matt-lipstick very sumptuously red for morning. She greets me in an elegant manner, graceful in her use of honorific and codified business speech.

An unsmiling guide from the personnel department arrives and we ascend to the fourth floor, I fear my person dissolving (all that identifies me as unique) as I follow obediently (in body and mind, reluctantly, as this recalls in part a culturally conditioned attitude) through a beige-colored metal door. The room is unexpectedly small (see Plate 1). The grey walls are yellowed from many decades of exposure to cigarette smoke. It is decorated sparingly with desiccated strips of ancient Sellotape, hanging from where once they had supported posters of some kind. Scaffolding against the large window obscures the external view. Along this window is a single row of three desks, beneath each of these desks emerges a double-row comprised of paired desks facing each other. Each double-row, the 12 desks, forms a unit; a section within the department. The executives sit at the three desks closest to the window. Then the most senior men of each section sit closest to them, and from there down younger men sit in descending order of age. Women occupy the desks nearest the door. The fax machine and two sets of photocopiers are placed close to the women, and tall, solid, metal cabinets and drawers frame the remaining space. Nothing seems out of place, the desk surfaces are sparse and neatly ordered. There are two meeting tables in this small room. One is used by the department head for briefing and by the two sections of the department closest to it. The other table is used by the remaining section for group meetings and dealing with large amounts of paper work.

The total number of employees at this head office is 314, of whom 86 are female and 228 are male. The majority of women are in their early twenties to late twenties (*dankai-junia-sedai*), whereas, for men, overall, there is a comparatively equal distribution of numbers throughout each cohort based on the year of entry. This head office is divided into 11 departments, each staffed by varying numbers of employees, and these departments are dispersed among 5 floors in the building. Each floor is under renovation. The office space is limited such that two or three departments would share the same floor, but some smaller departments are given their own rooms. This renovation is regarded positively by the workers who remark on how their former dismal surroundings failed to generate a sense of workplace pride. As soon as one floor is completed, it is quickly occupied to allow reformation of the next level. The delighted workers say this investment is long overdue, and indeed, it changed the working environment from officious sterility to one of warmth and brightness. The hard floors are carpeted in a shade of sky blue, larger desks with comfortable adjustable chairs are fitted, and large sliding cabinets are installed. The lavatory and tea rooms (*kyūtō-shitsu*) on each floor are transformed into brightly lit, pleasant spaces, with counter tops of soft speckled grey granite finish with fittings in baby-pink.

As a number of departments share the same floor, on average, 50 people potentially make use of the tea room, yet this tea room is barely large enough to accommodate 5 people at one time. Since the renovation, the tea room changed location. It is now set in an alcove along a corridor that leads out to the fire escape (see Plate 3). As the tea room is set back from the main corridor and work area it becomes a space that can only be observed when taking the staircase or attending the lavatory. However, pre-renovation, the tea room had been even more secluded from general view, as it was formerly located in the place now occupied by the male lavatory (post-renovation), and the tea room was larger and a door offered greater seclusion, but it has been removed.

Back in the main workspace, pre-renovation, the workers' own footsteps on the hard flooring made them self-conscious and aware of their relationship to others, and of their part in disturbing the others' concentration. The windows were small, and workers toiled in the artificial light of the fluorescent tubes reflecting off the cold floor and grey desks. Post-renovation, footfalls are muffled by carpeting and workers feel freed from the sense of others watching. The walls also gave way to the installation of larger windows, which contribute to a feeling of release from daily confinement, and to these I will often turn in the coming months. During these moments of escape, I feel myself part of the activity of the large metropolis, observing from multiple perspectives the other workers in office blocks opposite and the tiny figures below crossing the boulevard in waves, and above and afar, the turning of a concrete monotony to an urban beauty bathed in magnificent crimson sunsets.[2]

Why Study JCars?

Ethnographies about large Japanese workplaces are set in urban areas and most are set in the metropolitan area around Tokyo, perhaps mirroring the fact that to the Japanese, the city of Tokyo as an idea has represented the cultural and political core of Japan since the late nineteenth century (Smith 1978). In contrast, as a setting for an urban ethnography of the workplace, Osaka has been relatively unexplored. Osaka is Japan's second largest metropolitan area, and its culture—food, people, dialect, and sense of humor—is understood in juxtaposition to the style of life characterizing Tokyo. Osaka, according to my subjects, has affinity with passionate Latin cultures.

However, the reason that I chose a company in Osaka is due to the problem of limited access. Originally, I set out to study the identity of working-age returnees and applied in writing to 20 multinational companies (mainly based in Tokyo where the majority of returnees find employment), but found difficulty in gaining access to these large corporations—this experience is quite common among potential researchers regardless of one's provenance.[3] Many companies do not welcome intensive and potentially intrusive methods of qualitative research conducted over an extensive timeframe, partly due to unfamiliarity.[4] The research methodology applied in workplace studies differs substantially between Japan, Europe, and North America. As noted by Mouer and Kawanishi (2005: 24–8) on the whole, with some exceptions, Japanese researchers, mainly working under frameworks of sociology of work and industrial sociology, prefer to conduct macro studies using in-depth interviewing methodology and survey analysis, whereas European and North American researchers tend to use anthropological methodology, particularly participant observation. Furthermore, social anthropological research in Japan (by Japanese scholars based in Japan as opposed to Western institutions) studies rural culture or social problems in the city, rather than "studying up" among the elite and powerful (Nader 1969).

In my situation, my father was instrumental in gaining my placement. He has been an employee at JCars since graduating from university in the early 1970s.[5] It was through his negotiation skills, position, and connections to high status individuals (my father's close *senpai* [senior] from university heads the personnel department [Sato-*torishimari-yaku*]), that my unusual request was accepted. The family connection was essential to my access, but the process was by no means easy: my father had to negotiate with many colleagues (he does not speak of this; I heard through my mother).

A lesser known family fact, perhaps an unsurprising one, is that my mother also worked as an office lady at JCars in the early 1970s, at the time her brother-in-law (my uncle) worked in the personnel department, and his daughter (my cousin) later worked for the overseas department in the late 1980s. Although my father was working away from the head office (in the United States at the time), I had mixed feelings about conducting research at JCars. As researchers of unfamiliar social situations, we are all destined to commit unintended *faux pas,* and I didn't want to compromise my father's status and career success. However, I wished to experience and identify with what was important to my father's life.

Multiple Positions and Feelings

Participant observation[6] is a research method that immerses the anthropologist in an unfamiliar world, and this terrain of the unfamiliar includes spaces within the researcher's own society; it not only refers to exotic cultures. Anthropologists participate in the daily life and social relationships of the people they study and observe what they do and what they say about what they do in order to find out what is meaningful. But anthropologists also experience for themselves what it is like to live as their subjects do. Participating in this way, the researcher becomes positioned in particular ways in relation to their subjects. This means that ethnography is able to account for different levels of perspective in addition to displaying an ongoing negotiation of reality. The resulting interpretations are highly context-dependent and a commentary becomes useful. As the particular social and cultural context in which the anthropologist finds herself forms her subjectivity, this, too, when explained, elucidates how the social works. "To be hailed or addressed by a social interpellation is to be constituted discursively and socially at once" (Butler 1999: 120): thus in this section I show the multiple ways in which I found myself positioned in relation to my subjects, and the ways they worked together to shape my enquiry.

I began fieldwork as an intern (*kenshūsei*). Previous interns had been business studies students from North American, European, and East Asian universities. Therefore, my subjects did not refer to me as a researcher, although they were aware of my purpose to conduct participant observation and seemed impressed by its novelty; they saw me principally as an intern and student (*gakusei*) who was conducting research about them. My age, sex, and idiosyncrasies[7] might have contributed to this, but to me my position seemed appropriate as "intern" sounded less intrusive and politically neutral; in fact, both intern and student connote lower social status than workers (*shakaijin*). It seemed a befitting way to close the distance between my subjects and me.

I was also primarily identified in relation to my father, as the daughter (*o-jōsan*) of Kurihara-*shuseki* (head of an overseas subsidiary) who was well known to most workers at a personal level and if not, workers at least knew of him. Many male managers close to my father's peer group therefore acted with familiarity, affection, and informality toward me, which was in stark contrast to the distance and formality that characterized their relation to other interns. This was an advantage as I felt it was possible to approach the unapproachable, that is, men of high status, with questions more or less immediately. When interviewed, the younger managers of *shunin* and *kachō* class regularly dropped anecdotes of my father into our conversations; I appreciated their efforts to put me at ease in spite of the pressure they were under to perform well: phrases such as "you ask such difficult questions," "I'll have to come back to you on that," "that's deep," and "very interesting" peppered our dynamic sessions, and at the end the managers' faces relaxed into big smiles, relieved, their bodies noticeably freed of pressure, stretched out, momentarily for a few precious minutes almost lounging. However, with many others I had to build up rapport, as is the case with any researcher. Overall my association with my father had a triple function: it helped to foster feelings of familiarity among workers toward me; yet, it also increased the distance because he was of higher status and this placed me on the same cline or continuum as him, which needed to be downplayed on my part; and, bearing his name protected me to an unknown degree.

I was also defined in terms of my expatriate background: when I arrived at the office there was little need to explain that I had lived abroad for over half my life, and was indefinitely based in London—word had already got round. Japanese nationals who have lived abroad during their youth are known as returnees (*kikoku-shijo*); we are often reduced to Weberian ideal types, invested with symbolic value, that enable a discourse of "Japaneseness" among the Japanese (Goodman 1990). During the 1970s, when modernization drove the discourse of nationalism, theories of Japanese uniqueness was taken literally and returnees were perceived negatively: as different, presumptuous, opinionated, meddlesome, both incomplete Japanese and incomplete foreigners, unable to conform or readapt to Japanese society (Goodman 1990; White 1988). Yet, during the 1980s, when Japan embraced internationalization (*kokusaika*), returnees became associated with positive Western attributes and envied (Goodman 1990).

My subjects made sense of what they perceived to be my difference in manifold ways. Due to limited space I am unable to capture the rich nuance here and can only hope to be forgiven by the reader for simplifying but, to many women who did not share my background and to male managers who had never been abroad, I was a curiosity. In all senses: good, bad, in-between. To others, mainly young men who had no overseas experience, I was *shinsen* (refreshing) and admired in the sense of being seen to possess positive assets that the other could absorb and benefit from. They seemed impressed with the big project that I was undertaking. Two middle managers discussed me as though I wasn't party to their conversation; they defined me as "pure" (*puá*), "different" from Japanese women, I represented what Japanese women had given up. Their sentiment conveys a social commentary on the contemporary losses to patriarchy: a late thirties to early forties masculine lament toward the younger generation of Japanese women who are seen to have lost the traditional feminine qualities once symbolized by the ideal figure of the gentle, pure, and beautiful *yamato-nadeshiko*. Likewise, one man who perceived himself an outsider in Japanese society identified with me; and because he did, my difference to Japanese women, and the fact that I was a social researcher, was given positive qualities— sympathetic and genuine. Some men expressed envy of my situation, but they never turned their emotions against me as some women did. On a different note, a female worker, a few years older than me, in a joking tone, described me as *kegare* (a term used in reference to ritual pollution and disturbance of the social order).

Wada-*kachō*, a section-head, marks his authority to others present in the room by virtue of larynxal volume; this is deduced instantly by observing his telephone manner. On my first day, armed with a thin file containing my research proposal, his stride—sharp—punctured the frozen atmosphere. Launching swiftly into monologue, by which he detained the entire personnel department in aural captivity, he said, "neither the company nor I will tolerate inappropriate behavior." To reinforce his point, he quoted a proverb: *gō ni hamaru* (When in Rome, do as the Romans do). He added, "you shouldn't consider yourself exceptional for speaking fluent English." By no means am I an animated speaker, but, seeing my hand float to accompany my response, in a spectacularly conservative tone, he did not miss the opportunity to pounce, "what the hell was that?!" (*ima no nani?!*). Can silence be more penetrating? After he left the room, others were sympathetic. In consolation, the women in Wada's group instinctively vocalized their sympathy: they said he didn't own a passport, couldn't stand up to his wife, and that I shouldn't take much notice of him. In contrast, men became immobile, bodies rigid, necks retracted, they fixed their gaze low, seriously inspecting a spot on their computer screens. The senior managers including the head of the department were absent.

Approximately one month after the terms of my induction as an intern was spelled out, Wada, again, in front of the whole department, instructed a new recruit. This time a young man who entered mid-year following the collapse of his securities firm. Among other instructions, including "never arrive late for work" and "never call in sick no matter how hung-over you are from the previous nights' drinking," the new employee was directed, "always make sure you can speak English, train yourself to speak the minimum." To which Wada added, "although, I don't, because I hate it."

With another manager in the personnel department, as I addressed him in Japanese, all his responses were in English—this made me wonder whether interns and returnees are not meant to fit in, or was he compensating on my behalf? His handouts were meticulously folded and numbered by hand and he did not answer any of my questions, instead he just spoke about what he considered important. Was this a unique personal trait or managerial ruse?

There were four returnees (excluding me) working at JCars, who returned to attend Japanese universities, all of who were women. They experienced similar positions and feelings as those described above. Therefore, my personal experience was cultural. Ono-san from the overseas department had been working for six years. We met as children when we both lived in the United States and had spent several summer holidays playing together. Three years my senior by age, Ono introduced me to her colleagues in her department as well as to her friends within the head office as her childhood friend (*osana-najimi*). The word *osana-najimi* is more nuanced than "childhood friend" or "old friend," it implies a heartfelt communion cemented over many years; it originates from the Muromachi period and is derived from the word *najimu* (to be attached to) (*Nikkei* 21/11/06). Indeed, this term is often used with fondness and affection, and my association with Ono facilitated my integration into the office. She was certainly supportive, protective, and encouraging. Our style of relating was casual and informal. However, her perception of me was a farrago of inconsistencies mirroring the other workers' views of me, whereby she also referred to me as a foreigner (*gaijin*) and saw me as somewhat mysterious (*shinpi-teki*). This resulted in a shifting effect on the degrees of closeness we mutually felt for each other.

Overall, in the late 1990s the workers' perceptions of me in this office were often somewhere between the two poles of 1970s and 1980s sentiment described by Goodman above, and constantly shifting. The porosity of my subjectivity imposed a disorienting effect; I couldn't foresee its impact on my field study at the time.

Friendships, Worldview, Intersubjectivity, and Ethnography

My subjects were highly articulate, intelligent, witty, and at heart very kind and patient individuals who were enthusiastic to share their interpretations and experience of Japanese culture and the company. As one would expect to find in any office, this one contained a range of distinctive corporeal styles in behavior and appearance (indicating something about how they regard femininity and masculinity, and dominant notions of class that are tied to it). Feminine beauty takes a variety of forms; some women are vivacious and fun, while others are more serious and low-key. Some speak gently; some loudly. Some prefer to be heavily made up while some wear no make-up (just lipstick). Plump, skinny, short, tall, pale, or with sun-kissed skin. As for my feelings toward my subjects, these shifted variously as much as my subject's perceptions of me were prone to fluctuation. I experienced greater degrees of shifting in my feelings toward female workers than with male workers. This is because I considered some women in the office as my "friends," identifiable on account of our common interests, experiences of Western culture (in some cases our earlier childhood friendships), our sense of humor, or based on a neutral and reciprocal mutual interest in and affection for one another. In such friendships, trust was present and mutual exchange of private thoughts and feelings took place openly. Furthermore, there was no leakage of information to other parties, at least none that came back with negative consequences.

Friendship has the ring of egalitarianism and innocuousness about it, but some friendships revealed a definite utilitarian component within the relationship. Being on friendly terms set up a chain of expectations where, periodically, strong pressure built up around the question of what I could do for them (cf., Gilligan 2003: 59–60). My mixed reaction to their expectations led me to question "friendships" as a category of interaction in working life (cf., Hendry 1992). To be fair, I had entered into relations with them with expectations of my own, that is, to understand how corporate discourse shapes, and in turn is shaped by, the lives of workers. Perhaps, in the face of my unending inquisitiveness, by being themselves, my subjects felt they were upholding their part in the exchange. Thus, it would only be right that I give back in return. Only fieldwork places the researcher and researched on unequal footing in this regard. But, then, I did wonder how the situation would have played out had I been positioned differently in relation to them, for instance had I been (considerably) older than my subjects or had I been of a different race or nationality.

Given that secrecy/privacy and inconsistent flows of information are grounded as by-products of social distancing (unintentional or otherwise) between colleagues, this is not a characteristic of Japanese workplaces as the same can be said of British workplaces; and knowing that I was able and willing to speak to everyone, passing fluidly between cliques and departments, many women who did not share my unusual position wanted me to spill the gossip on their colleagues. I was often probed for information about the men I interviewed, what it was like to socialize with a certain group or individual, skeptically whether or not I found so-and-so interesting, what does my *senpai* think about me, why is she still here and what happened to her marriage plans, and various questions of this ilk. I felt uncomfortable in having to manage the various expectations of my subjects that I felt at times were mobilized by malign motives; it was especially difficult while trying to assume neutrality (unachievable in any case) and faced with caveats, such as "don't tell x, but you can tell y and z." In certain situations, "friendship" was extended as a bartering tool in attempts to extract information. These women revealed enough as an offering of friendship but some facts did not fit or did not feel right and, in hindsight, it was apparent that their real motives had been concealed. I noticed this demand for gossip among interactions with a minority of women who (1) claimed authority in positions vis-à-vis me; and (2) put on a warm egalitarian front, the veil of friendship, and maintained their distance with this polite exterior. With the first type of relationship, I was being caught up in hierarchical relations of obligation and expectation. Whereas, in the second, I was witnessing the distinction of *omote* (face, appearances, surface) and *ura* (mind, beneath the surface)—two aspects of a single entity (Bachnik 1998; Doi 1985). The two types can coexist in one relationship as well. The shadow that lay behind relationships emerges, a constant reminder that relationships are always embedded within power differences and status politics. Withholding information induced guilt feelings, even though it was correct and responsible to do so for all concerned; this guilt precipitated from my empathy toward their desire to know, that which Aristotle says is inherent to human nature, the accompanying fear of causing antagonism, and my desire to be friends.

Vincent Crapanzano (1986: 51–2) likens the ethnographer at the writer's desk to Hermes: both are the interpreter (of the foreign culture), the marker of boundaries (between authorship and readership), the stealthy messenger, and trickster (using rhetorical tools). I push this analogy and add that, while anthropologists are in the field, their subjects look to them for gifts, as they might to Hermes, the bringer of dreams, the god who unites strangers across shadowy boundaries. Sadly, anthropologists do not possess golden winged sandals, only Marc Jacobs ballerinas, and wishes all too often remain unfulfilled. And, in this instance, like the Maussian (1990) concept of the gift, a break in the flow of reciprocity dilutes bonds of solidarity.

During research we encounter this problem of (not) meeting the expectations of our subjects: Karen Kelsky (2001) shows in her fascinating transnational study of desire (*akogare*), how her subjects, a class of internationalist Japanese women, assumed that Kelsky should share the same dualist worldview they hold of Japan-West relations; consisting of (in brief) a Japan characterized as a backward patriarchal, corporate, and familial system deserving of criticism in contrast to the West, liberal, and liberating Oriental women. Kelsky's feminist perspective and marriage to a Japanese man (which transgressed her subject's unspoken code) constrained rapport building between her and her female subjects; one relation moved from friendship to estrangement, which vexed Kelsky. Furthermore, white men put her in a therapist's role and bombarded her with confessions about their sexual involvement with Japanese women via email and telephone (pp. 237–47). She predicts, rightly, that anthropologists researching discursive formations that unfold in transnational spaces will be required to respond to unanticipated encounters such as these more frequently.

At one time feelings ran high. Here I wish to convey how a given individual's embodiment displays signs of difference that impinge upon friendships and impact research practice. Strong emotions are revealing: such disruptions highlight cultural boundaries and material conditions, thereby showing us that personal experiences link directly to sites of production of cultural knowledge. It was early June, around the time I returned from a month's break in London (to gain distance from the field and gather my thoughts). Many women were nice to me and said they'd missed me. In fact, I had kept in regular contact with approximately 10 women by email during my month away (including another returnee from my lunch group). The conversation in my lunch group was much more dynamic and inclusive to my presence; I no longer felt literally on the periphery, and a few women who until then had never looked me in the eyes or addressed me directly in a conversation were doing so now. Paradoxically, these very same women became openly hostile as well. It seemed that my withdrawal from the office by going to London had a large part to play in expediting strong feelings of envy, which until then had been expressed weakly in the occasional comment. In the time that I was away, it could be that my subjects had processed their feelings about me, for there was nothing to suggest that a change had taken place in the external situation of the firm that would have prompted some of my subjects to become more resentful of outsiders.

This disturbing period brimming with projections of envy and dissatisfaction, verbal and nonverbal, lasted about one month. Days continued where I did not want to go to the office. These women's comments juxtaposed our mutual advantages and disadvantages, centering on an evaluation of lack, specifically theirs, which my presence appears to have exposed, such as the opportunity to pursue postgraduate study abroad and the freedom implicit to going abroad; a parent's capacity and willingness to finance a postgraduate education overseas; and the amount of work assigned—the fact that my job responsibilities at JCars were lighter than theirs. Alternatively, the hostility could be read as an expression of widening socio-economic inequality, manifested here in the form of increased competitiveness among women. What was I to make of the fact that they disliked the way I differed from them, a fact that I cannot hide since we embody our experiences, emotion, thoughts, and history (Bourdieu 1990: 54–6, 73)? Having experienced childhood in Japan between ages 11 and 15 in the 1980s as a returnee, and now faced with similar reactions in the workplace, I knew that what seemed like personal attacks had origins elsewhere. To follow the psychoanalyst, Nancy Chodorow (1999), we see in this exchange of emotions how cultural meaning and personal meaning are inextricably fused; feelings are culturally determined but also powerfully experienced in our psyche and this orients our relations to others.

As many researchers agree, the personal or the private and the social are interlinked and inseparable, and this experience serves to reinforce the earlier point on ethnographic writing that we cannot omit personal experiences from our ethnography. I endured the arrows, decoded them after a time, and sometimes fired back. Although it is impossible to verify, I think my situation could have been far worse had I not had my father's name to bear, his presence in the background, this invaluable shield that protected me by keeping my subjects' behaviors in check.

Let us remind ourselves that ethnographic research is an intersubjective project, the analysis arises from a specific cultural encounter, the nature of which therefore requires some explanation. Two important themes emerge through reflecting on my own emotions. First is that the value of friendship is not so much a reflection of differences between Western and Japanese criteria of friendship as a reflection of the nature of contingent or office relationships (cf., Hendry 1992). For this reason the variations in the patterns of communication that sustained office relationships, particularly friendships, became interesting to me. Second is the interdependent question of subjectivity and friendship—why my subjects focus on my difference and define my subjectivity in ways described above, and conversely, why, at times, I see some of my subjects in the crudest sense, as a human being reduced to a Dostoevskean "principle." Is it simply due to a tendency to generalize from disappointments? Perhaps. But, in fact, this points to the link between friendships, subjectivity, intersubjectivity, and worldview.

My sentiments certainly reflect unfulfilled desires (not expectation) of my own—not to be trapped by a symbolic identity imposed by the other, and simultaneously to be accepted as a worker and as a Japanese national (although I was unable to officially belong in an employee's sense because I was a fieldworker). The most burdensome element to this persistent exteriorization, inescapable and indefatigable, was the dual structure of criticism: criticized for my ambiguous presence (having more than one meaning or culture in this case) *and* denied a status of ambivalence (to have mixed feelings or cultures coexisting).[8] This is an impossible position.

Just as every social interrelation allows us to reflect on our experience of society, our subject's worldview can be witnessed in their view of the anthropologist. Their perceptions of this anthropologist derive in part from their orientation to Japanese society and corporate culture—or corporate-centered society—that is structured by competing discourses of tradition and freedom. My subjects themselves are at odds with the interplay of forces shaping their world. My temporary presence among them stirred up, and I became witness to complex feelings around agency, desire, and social control (tradition). Other returnees were also sensitive to this feeling.

At a philosophical level that speaks to us all, in thinking about this dynamic, we recall Martin Heidegger's (1962) concern with the question of being. Our being-in-the-world is structured by the tension between being true to ourselves in deciding our own course of life, and, conforming to the world with its constraints, thereby surrendering to other people. With respect to Japan, many writers have commented on the tension shaping women's lives. To cite one example, Nancy Rosenberger (1995) shows how unmarried women in their early thirties are caught up in a duty to conform to the modest, nurturing housewife role rooted in conservative Japanese family and corporate ideals. While individuated by their role as consumers, women can fulfill "selfish" desires and indulge in leisure, travel, and shopping, and thereby invest the body with sexuality, sophistication, and a global identity. It is true that in the past 20 years Japanese women's life course has diversified (Miura 2005): a woman today has more choices, she can become a housewife (full-time or working part-time), continue to work, become a *freeter*, an agency contract worker, delay marriage, stay single, have children outside of marriage. Yet, this diversification does not apply to all women. Living in an open cage inhibits women's views and alters their experience of the world; the anticipation of limited options recycles tradition, thereby maintaining its position as the ideal. Transformation becomes illusory. The ensuing curtailment of the enthusiasm necessary to go out into the world perhaps explains why tradition is endorsed more effectively among the class of women who work at JCars. This is also true of its male employees.

Echoing Rosenberger's observations, both my male and female subjects voiced tension between traditional and alternative ways of working and living. These feelings are pronounced particularly because Japan is in a stage of transition. How do they negotiate this plurality within the system? Let us take the example of unmarried, 20-something women at JCars. I should surmise that these examples speak to this class of elite workers within the wider society, although obviously I do not speak for every woman. Strict parental (i.e., cultural) supervision enforces a 10PM curfew because women generally live at home until they marry. In order to maintain some freedom to conduct their love affairs women resort to double lives, the deception here functions as rebellion against parents who control their sexuality. I appreciate that this level of parental control may strike some readers as rather draconian and quite extraordinary, but I only note the knowledge gained from my subjects by hearing about and standing witness to their lives.

Some women seek traditional arranged marriages to doctors and dentists that guarantee long-term financial security, social prestige, and comfort in spite of, or rather because of, patriarchy. In the short-term, the prelude to an arranged marriage provides the Prada bags, expensive dinners, Royal Doulton wedding china, and wedding ceremony at the most exclusive venue, Teikoku Hotel. Yet, such women willing to marry for wealth also desire passion and romance. The two are separate, although both are intensely commodified. Such women do not expect to find romantic love or passion from arranged marriages (Applbaum 1995). It is a matter of course that they must desire and obtain both simultaneously, albeit from different sources. Alternatively, to fill the years until their arranged marriage, women might date freely for fun and romance with no view of commitment.

Others experience severe job and life dissatisfaction, and escape the conservative family, company, and cultural norms and rules through study or work abroad (cf., Kelsky 2001). This solution proves to be temporary and incomplete as their individuating actions are negated once they return, when they have no choice but to fit in with societal rules. There are women who are not permitted to relax after returning home from the stressful office day. They are expected to tend to their parents' and grandparents' needs and feelings above their own. Tearful and frustrated, always wanting something other while adapting to constraints, always somewhere in-between. Under these conditions, many women cannot experience a true period of singlehood in their lives, nor can they be true to themselves. These realities display the variety of ways women shape their lives while necessarily surrendering to others and to cultural norms that structure their worlds.

Furthermore, there is still a wider context of meaning in which I would situate my subjects' worldviews sketched above and my subjectivity formed in this fieldwork encounter, these being interrelated. Still it involves transcending dualities. I question the meaning of living in a globalized, cosmopolitan world when individuals, not only women or returnees, are nonetheless still expected to conform to essentialist representations, difference—more often than not—being something negative that needs ironing out. On this, I make two points:

(1) Some readers may wonder, quite reasonably, how we might account for the view that Japan is more tolerant of diversity and nonconformity, especially since the late 1990s (Kingston 2004; Nathan 2004). In response, I should like to stress that the problems of difference and othering and a greater tolerance for diversity in society are not mutually exclusive. Necessarily I recall my emphasis on the dialectical tension between coexisting categories. As the following chapters show, this book documents the recent social transitions toward greater acceptance of diversity, but, new forms of identity, work, of relating to others unfold in tension with the ideology or tradition they contest. Let me strike another reminder, that anthropology's understanding of culture is created within a specific set of encounters and this leaves its traces on the ethnographic text: (a) What we think of as knowledge oscillates when we examine the meaning of situations and interactions at close range, then at long range, and we constantly shuttle between the two. For example, if a person is sad because of something bad happening to them, this may leave them in a tender emotional state temporarily, but we would not necessarily conclude from this observation that the person is a depressive character or that bad things happen to that person consistently. The same can be said of accounts of the workplace in relation to societal change. (b) The nature of anthropological knowledge is intersubjective. From my field research conducted at an elite corporation, I interpret what I experienced, and in this context, it is the tension between ideology and creativity that this account explains in depth. To be sure, we can be optimistic for the future of employment practices and toward a better understanding and acceptance of gender issues and other forms of difference, such as disability, but let us not forget that there exist multiple perspectives and contexts: it depends on the subject and their situation, when and how you pose questions; and, who you are; then on who people think you are; and on where one looks, center and margin, the surface and concealments, situations that allow for greater acceptance of diversity and those that do not. Aspects of "Japan" will be experienced very differently depending on the individual's gender (man, woman); sexuality (homo-, hetero- or asexual); age (infant, child, adolescent, middle-age, elderly); level of education (high school, junior college, university, vocational college); occupation (white- or blue-collar) and size and reputation of the organization; place of residence; nationality and immigrant status, and so on. Effectively, there are multiple ways of analyzing and knowing "Japan." The purpose of anthropology is to make these differences speak, not mute.

(2) I am also attempting to convey that Japan is not unique to the burden of othering. Although I speak principally from my position as a fieldworker in Japan, at a given time, at a given situation, when I write that othering is a

profound social problem, I am also speaking as a long-term resident in London where I encounter a different set of social barriers to do with race, gender, and class on a daily basis. Our contemporary cosmopolitan world presents a paradox for subject positionings because, our reality consists of plural worlds and while each communicates with the other, self-identity is increasingly formed and articulated using an image of the other (Augé 1999). This perennial problem reverberates throughout history and across cultures.

Some comfort might be gained by invoking the description of the "foreigner" who manifests contrary characteristics—on physical, psychological, emotional levels—that coalesce into a universal figure, in whom, Julia Kristeva (1991) claims, everyone can recognize themselves. If we are all foreigners in some aspect and degree, if people identified with the otherness within themselves as Kristeva suggests, and if people consciously exercise an aware subjectivity, could this begin to resolve the problem of othering and dilute the persistent urge to structure and simplify the world in dualist terms?

We might recall that we also learnt from Georg Simmel (1971) that the stranger occupies a paradoxical position, geographically nearby but psychologically and culturally remote. Where Kristeva's foreigner shoulders the melancholy that stems from psychoanalytical forces of separation and loss, which she describes poignantly with accuracy, it is freedom that distinguishes Simmel's stranger, who, from a privileged position, as a receptacle of secrets, participates in society from an objective standpoint, which is considered to be both positive and definite.

My fieldwork at JCars was certainly very interesting and my subjects were caring individuals. I highlight these problems in order to demonstrate the ways we might make our encounters in fieldwork and subsequent analyses more productive. Every anthropologist will tell you that fieldwork is difficult. If it is easy there is little point doing it. I do wonder, however, despite the personal and cultural being intimately linked in the ethnographic encounter, whether I could have maintained better emotional boundaries between myself and some (not all) of my subjects. Perhaps then I would have felt less vulnerable. However, one cannot expect others to share their lives and emotions without exposing our own in return. As for the relevance of personal experience to anthropology? Surely, it is that our greater vulnerability (see Behar 1996) heightens our awareness to the ways various knowledges and social practices are shaped. It is a bitter and invaluable pill. This burden is temporary and a privilege.

After fieldwork comes writing. Further to the preceding discussion of my positioning in the field let us ponder the question of positioning in the written text. Positioning, in this instance, is a vehicle that enables us to face our honest feelings about the fieldwork situation and our relation to our subjects; it clarifies the basis of the cultural knowledge produced and conveyed in the ethnography. In this, somewhat invariably, we might, rather *I*, certainly, identify in my self-presentation as anthropologist, infrequently or inconstantly but nonetheless so, a donning of a mask, as tender victim. Is this mask, we should wonder, a psychological trope that compensates for a privileged identity as observer and commentator on people's worlds and their intimate thoughts, an anthropologist's atonement for the role as portraitist who fixes subjects in time, mute? Postmodern argument on ethnographic writing attributes equal power to subjects and writers, as subjects are better able to respond to anthropologist's representations, but how adequately have I as well as others responded to this call (see Kapferer 1988: 99)? In addition, empathy transforms to guilt when we internalize our subject's feelings of envy and jealous behavior. Why guilt? Perhaps because simply by being there this anthropologist set in motion reactions in my subjects that evidence a possible dissolution of an illusion of freedom and independence or whatever it may be that hitherto made social constraints bearable. And, more generally, as writers, do we not attempt to escape the uncomfortable subject-object confrontation that accompanies the telling of the story? In summary, we tread a fine line: our text enables us to capture accurately the moment (however fleeting or semi-fixed) when we were objects to our subjects' discourses, but in our account of this we might (unconsciously) be attempting to redress an imbalance external to the question at hand. There is no prescribed way in which writers experience their relation to their subjects and the text, but what we can say is this: we have a responsibility to at least point our awareness at this process, as we, after all, write about the lives of others, this writing that rises from within us.

Techniques of Data Collection

The project of ethnography is to make visible the lives of women and men with whom the anthropologist spent time and got to know in depth. Early on in my study, I began collecting paperwork, the company documents ranging from those that show Human Resources policies detailing recommendations and implementation, to corporate Public Relations documents, and JCars newsletters for internal and external circulation, newspaper clippings concerning labor markets, individual life and career histories, and dreams and hopes for the future. As I spent more time with my subjects, interesting topics of conversation emerged spontaneously, such as, in brief, details of home lives, love lives, night lives, sex lives, what they eat, where they go to eat out, fashion, their beauty regimen—including hair depilation, eyelash growth promoters, and aesthetic treatments, make-up, diets— where they shop, domestic and foreign travel, contemporary music and concerts (*raibu*), movies and art, and sport.

Not only did I have a highly participatory role, assuming real responsibility for the tasks performed, but I also observed routines performed at work to understand how work gets done, and to know which social relations are important to this. I observed the level of differences at which relationships are formally and informally structured and of how they reinforce as well as work against the other, and the variation in content of interactions at the two levels. I noted discursive and behavioral cues, especially repetitions that show up underlying rules and structures, taking care to distinguish between appearances and reality, and who was saying what, when and how, and the effects produced, the value judgments. I also observed official events, such as the induction seminar for new recruits and office parties, and more relaxed social events. I also observed the way different people's bodies negotiate spaces—open and hidden spaces, gendered spaces, the Internet space.

In workplace studies we must not omit the lives our subjects lead in venues outside of the office: Marc Augé (1999: 94–5) (quoting Althabe [1990]) reminds us that information gained in formal interviews, meetings, and through simple observation ("events of communication") is the subject's self-representation, this is always partial, because something "private" is withheld from the researcher and kept for the subjects themselves. A fuller picture emerges by looking into (1) how subjects are shaped within the multiple spaces they pass through every day, such as markets, bars, and cafés; (2) these "spaces of communication" or "'elsewheres' in relation to the content they put at a distance: the city, the family, work" (ibid.).

Many of my subjects preferred not to be tape recorded; cautious of the consequences of leaving recorded evidence, but they were not averse to note-taking on my part, regarding this technique as more scholarly and less threatening. I noted the conversations verbatim to the best of my ability in Japanese, supplementing these with notes and interpretations in English and to a lesser degree in Japanese. Interviews, on average, lasted a minimum of one hour and extended to two hours depending on the individual and the topic.

My aim was to interact with as many workers as possible, but my subject's age, sex, degree of curiosity, and sociability affected my sample. Also I became acquainted far more intensely with workers of my section in the personnel and overseas departments. No one declined my interview requests, but some subjects were noticeably overcome by shyness and were tongue-tied. However, over the course of repeat interviews, these individuals overcame their shyness.

Inside the Office

I found my subjects' incessant commenting on my difference rather irksome, which has its history, as described previously. On my yearly visit back to Japan I became hypercritical of my own form, movement, and gestures (cf., Kondo 1986) and what Mauss (1979) calls "techniques of the body" that vary between societies. From the start, therefore, I used my body as an effective research tool to draw out definitions of "normal" in this corporate setting. This method guaranteed that I keep my distance from workplace culture as well. I deliberately deviated from the ideals of femininity of the female office worker; effectively, I made semi-concrete the differences and boundaries existing between my subjects and myself, but implicitly enough so that I could still participate and observe in their milieu. Eschewing the sophisticated look, I got a boyish haircut that I adorned sometimes with little butterfly clips, wore kitsch rings preferred by teenagers—my favorite at the time was a little pink plastic strawberry (while other women wore Tiffany rings)—and colored my nails in rouge noir rather than in neutral shades of pale pink and nude beige. One subject labeled my style *mōdo-kei* (wearing trendy and simple shapes in monotone) while others said, and not as a compliment, that I looked like a high school student.

During the initial stages of fieldwork I was given a seat in the personnel department. I went there although my father had arranged for the overseas department to take me on. This was because Sato-*torishimari-yaku*, the head of the personnel department, thought the members of the overseas department would give me too much real work to do, and I would be better placed to do my research in his department. In the personnel department there was complete silence and little movement and if there was, it was very constrained. The atmosphere was tense and stifled. There was no laughter. The character of the department matches its elite reputation, derived from their indirect involvement in corporate governance as they develop future leaders and board members, and directly through the head of department sitting on the board of directors (Jacoby 2007: 7).

My provisional role was to assist the group in charge of training the employees going abroad and interns coming to Japan, but nothing came of this. I was restricted in the research I was able to do as the managers seemed to have some interest in controlling my autonomy. This may have been in part because they didn't fully understand my research methods; the rest is political. Although I felt constrained, I began by observing and taking notes everywhere and at any time. I asked to see their collection of management journals and newspapers and used the Internet for my research to find returnees in the Kansai area in preparation for interviews. Taniguchi-*kachō*, a section head in the department, helped me to find returnees at companies in Osaka through his business connections. I also began formal interviews with returnees and other interns at JCars. Taniguchi took the first step in arranging these interviews by speaking to the section head of the interviewee's work group, and after the first few sessions I took over the organization from him. I would phone the individual and arrange interview sessions (Sunderland 1999). It was inappropriate to talk openly at length in an open plan office so interviews were held in one of the conference rooms by reception on the ground floor, which were usually reserved for meeting clients. Also in the reception area I sometimes used the tables in the open space. The tables are separated by high partitions, although these offer less privacy than the individual rooms, or we would interview in the subject's department. As the topic of my research changed and I became more confident, I began to interview more widely, from as many departments as possible. Overall, the frequency of interviewing fluctuated throughout the day, week, or month. On any one day I would interview between 1 and 4 workers or any number between 2 and 10 subjects a week, either formally or informally. I usually prepared a list of questions, and judged how to proceed throughout the course of the session.

My participation in the personnel department's activities intensified notably around the time of the new recruits seminar in April, and I got to know the members quite well. When I moved to the overseas department in July I was able to come and go freely between the two departments. I spoke to the managers in the personnel department on an individual basis mainly during interviews. These managers were each responsible for separate personnel issues, so it was necessary to speak to them regularly. They readily answered my inquiries in depth, often supplying me with official quantitative data and, using the time in these interviews, we talked about Japanese society, the company and about family and interests.

On the cusp of changing departments, I made a mistake. This incident illustrates the impossibility of being both inside and outside at the same time. A board member in the overseas department, Kimura-*jōmu*, was casually chatting to me and asked how things were going. Matter-of-factly I mentioned that I was waiting to get permission for a trip to Tokyo, but things were held up. He wanted to know who was holding it up and wanted to intervene. He suggested that we go together to speak to Taniguchi, my manager in the personnel department. I said I would speak to him again on my own. Exactly one week later Kimura came looking for me. He wanted to know whether my trip had been approved. I had not heard back from Taniguchi. This gave an opportunity for Kimura to express disapproval of Taniguchi's character, saying that he didn't know what kind of man he was. Although I had an inkling that it would be wrong to have a very senior individual intervene on my behalf, I was an intern after all, Kimura genuinely seemed to want to help, and I felt that it would be effective if he did. I was short on time.

That day Kimura spoke to Sato (head of personnel) and my trip was approved. Sato said that in future whenever I had similar requests I could ask him directly. Taniguchi's position was compromised in this situation because he had sat on my request far too long. From what Kimura told me, the incident identified faulty lines of communication between Taniguchi and his boss, Sato. But I am not sure if and to what extent Taniguchi was reprimanded on my account. I apologized to Taniguchi by explaining that I did not deliberately intend to go around him; it was accepted. When I mentioned this incident to Kubo-san, a male worker in his early thirties, he thought that I should have waited and not by-passed hierarchy no matter how slow my manager's response. The ordering of status among lower ranks did not matter to Kimura because he was at the top. It also showed that information travels fast at the top level. I learned that even while, or especially due to, having connections at the top, I would have to mind my position in the hierarchy, particularly, as an intern is among the bottom ranks. After this incident, I became intrigued by aspects of hierarchy as an investigative theme.

I moved to the overseas department mid-way into my research (end of May 1998). Ono-san engineered this as well as the managers in the department through our mutual consent. This department was light, airy, and bustling with activity; more women, more noise, more snacks. The ambience—a soft chorus of murmuring fax machines, whirring photocopiers and purring telephones—was soothing.

I was given official work, the same as a full-time worker, whereas in the personnel department I was not. I sorted and delivered newsletters and circulars to the designated pigeonholes of overseas subsidiaries; photocopied and labeled multiple sets of documents; shared in the cleaning duties of the tea room; processed company data of standard lead times for the production administration department; managed the accounts of monthly air shipments and ship-back documents using the Excel package; translated official documents, letters, and facsimile messages into English; participated in some departmental and section meetings; and edited the company's English newsletter.

Men of various ranks from manager to senior executives in the overseas department who knew my father spoke to me casually during office hours. Typically, such men had worked in posts abroad and their children had similar backgrounds to me. We would speak about my research (one board member was very learned in anthropology), current affairs, boyfriends, cultural differences, and gender in Japan. Three managers in particular regularly brought me newspaper articles on women and employment, and if I could not find a book I needed they would call the Tokyo office and arrange to get it to me. One manager gave me a CD of 1970s Korean pop songs, which I think was given by a client. People were certainly more approachable in this department.

Invariably, I spent a greater amount of time with women. There was a gender-based division in the organization of lunch groups based on the year of entry (*dōki*) and, as I mentioned above, I was allocated to a group of 20 women who had most recently entered the company. I ate with this group frequently and became well acquainted with its members, although I was also invited to join other female lunch groups. I would invite and be invited to eat lunch outside of the office in nearby cafés and restaurants with various women as well, either individually or in small groups of three to five members. I had casual conversations with women in the tea room and changing room where women would go off during work hours for unofficial breaks, and in passing along corridors or through emailing (email was connected six months before the start of my fieldwork).

At times I ate on my own during lunch, both out of choice and by necessity. I also sat at my desk among a few men who were eating packed lunches ordered by the company (*kyūshoku*) or brought from home (*o-bentō*). I often conversed with the manager seated to my left and opposite me on these occasions or to whoever came along. If no one close by was eating lunch at their desk, I would read my book, write in my notebook, write letters and emails to friends in London.

Outside the Office

Official Contexts

In Japanese organizations, including schools and companies, parties or *enkai* are held within an allocated budget, on occasions such as workers leaving (*kansōkai* or *sōbetsukai*) or entering the group (*kangeikai*), at times these being combined (*sōgeikai*), and at the beginning (*shinnenkai*) and end (*bōnenkai*) of the year. These parties could be both formal or informal depending on the character of the department and the members of management present. For example with board members presiding, the occasion became more formal. But, in this work-related context, I was still able to speak to all members on a relatively informal note because the topics of conversation exchanged were casual despite the formal atmosphere. And usually as the party progressed, with a modicum of alcohol, the workers became more talkative, spontaneous, and free in their conversations.

Casual Contexts

On a few occasions, managers who were close to my father through having gone to the same university or having worked together took me and other returnees from their section out for supper or lunch at weekends. Most often these occasions were casual as I related to these managers not as an intern, colleague, or senior worker, but as a family friend. For the manager this was part obligation, part social. They seemed to enjoy our company and took pleasure in the fatherly role of taking us to eat lovely food in nice places. Interaction with younger men took place in the context of casually organized suppers in *dōki* groups comprised of men and women after work (*dōki-kai*), or at weekends.

I also went along to parties (*nomi-kai*) organized through Ono-san who, through her university connections, brought together men from different organizations. Selecting participants from a network of co-workers or class-mates functions to guarantee their character and social background. This type of dating party has a different atmosphere to the suppers organized in the *dōki* group. They involve more flirting because it was purposely organized to create couples. This party or *konpa* is held exclusively for and among young unmarried people. On these occasions, as some people are meeting for the first time, it was important that a gregarious and humorous individual take on the role of party supervisor (*mori-age-yaku*). This role falls to men and they work hard to boost the party atmosphere. I found it very interesting that a casual get-together would involve such carefully choreographed role assignment. Making the encounter seem casual perhaps requires a lot of effort, particularly if the party is to be a success.

I saw some of the women in my section and those from my fellow *dōki* group at weekends. We would go on tourist excursions to nearby cities, to see pop bands, window-shop or to relax, eat, drink, and chat. I went for supper after work with women whom I knew less compared to those with whom I spent the occasional weekends.

My time was also spent with neighbors, old school friends from the time I lived in Osaka and Nara, and Ministry of Education–funded English language junior high school and high school teachers (English, American, and Canadian graduates).

Email in Research Settings: As Method and Topic

When I arrived at the office, and throughout my fieldwork, I noticed with some curiosity the manner in which workers would often hide their com-puter screens by physically shielding them from my view or by bringing up a work-related document a fraction of a second too late as I passed behind their seats. As I settled into my department, I witnessed a prolific exchange of email among women for personal rather than for work purposes. This practice appeared to be less prevalent among men (in managerial and junior positions). Therefore, I identified this as a gendered practice that was rela-tively specific to female workers.

In order to learn more about email practices, I wanted to participate in this communication. This was not straightforward, as interns had no access to personal laptops. In order to send and receive emails I used the free-standing desktop PC (not laptop). To do this I had to leave my desk, walk to the PC, and wait to be connected by a slow modem (see Plate 3) then access my external account on the Hotmail website. All this causes a time lag and the momentum of the conversation is lost to me. Whereas the workers need not (could not in any case) log on to the external Internet to check for in-coming email, they could email each other in real time. I was reluctant to inform my manager of my interest in studying the social aspects of the uses of personal email. It could compromise the female workers if the manager asked to review my progress. Further, I was concerned that the women would be less open. However, in keeping with fieldwork ethics, I did tell my key female subjects about my interests and they appeared not to be concerned that I took notes on this topic. I was using the computer linked to the Internet for my translation and administrative work for JCars: Managers in my department would often get up to use the photocopier placed behind this computer, which meant the route passed closely behind me. This was a natural route taken through the space of the office and it was not a deliberate tactic of surveillance on their part. Given these circumstances data on email is largely based on interviews, observations, and what I picked up from daily conversations.

Email, then, is both a topic and method of research. Email has continued to be a means to follow-up my research since returning from fieldwork. It has been a way of rechecking data for accuracy and to find further details about initial observations. Further, email facilitated the two-way exchange of information; as researchers, we need to participate in reciprocal communication.

Summary of Methodology

Japanese workplaces are seen as male domains (Abegglen 1958; Clark 1979; Cole 1971; Kondo 1990a; Lebra 1981, 1984; Lo 1990; Nakane 1970; Ogasawara 1998; Rohlen 1974; Vogel 1963). Through my position as a young, female intern, researcher, daughter of an employee, and returnee, I found that access to people, even older male managers was not a major hurdle. If women really are the outsiders that they are purported to be in the literature on Japanese workplaces, I think I would have found access on the whole much more difficult. Gender is one of many factors that underpin outsider and insider status. Through my experience I began to see that women's status in the workplace can be conceptualized differently, and that a stringent bifurcation of social life need not characterize workplace ethnographies. I did not entirely escape the gender role ascription, but, when possible, I chose when and when not to conform, and I think this was the same for my subjects. Furthermore, which of my positions functioned at the forefront to define how I was seen by people was always shifting. Although it seemed to me that gender did not pose serious limitations to my research, most of my close contacts were women, but this was partly a personal preference as, generally, I find relating to women easier than relating to men. I think this allowed me to write about women's relationships to other women in depth. These women's interrelationships are analyzed in relation to women's workplace relations with men. Instead of overemphasizing the latter in studying status politics, I sought to inform the reader with a balanced perspective.

However, it is likely there are things that I missed by being a woman and having a female perspective, such as participation in certain events that therefore do not figure in this ethnography. For example, I did not have access to places such as the male dormitory. In addition, I encountered problems over issues of access principally with regard to certain types of information. Information considered private or contentious was naturally difficult to obtain, such as aspects of company policy that remained confidential. But over a period of 12 months as I developed and maintained relations with a well-rounded range of subjects varying by age and gender, who on account of their position had differing levels of access to different kinds of knowledge, I think I was able to obtain sufficient information on which to base my fieldwork findings and analysis.

I have shown in this chapter how the ideas that shape this ethnography are due to a combination of my situatedness and theoretical interests. The events unfolding in the workplace at the time of my fieldwork, for example, economic restructuring, the renovation of the office, the installation of interactive technology, and my relationship to my subjects as friends, led me to explore these issues through the various ways of how the management discourse shapes its notion of community, time, space, and social relationships. During data collection, I was unsure what form the ethnography would take or how my subjects would appear in the narrative. I only gave thought to this when I began writing; and I decided to focus on conveying immediacy and authenticity—in other words, to tell it like it was, to convey office life in all its richness. I think any other anthropologist in my position would have done the same. Now with thoughts on my relation to the field and on shaping this ethnography I want to intimate a connection between the anthropological endeavor and a "liaison":

> What had begun as casual and flirtatious was acquiring purpose...as if whilst simply and hedonistically enjoying the company of the other, they had each in their gradual submission inadvertently developed—alongside intimacy—an emotional requirement, and shockingly...a dependency...A bonding...of pleasures, wants, of common reference and anecdotes, of shared memory and of language...They had become inseparable—their reciprocal affection becoming almost a need...then to establish itself into a pattern of exchange. The more of themselves they gave to the other, then in part to reside there, so the more they sought the other out, that they might there find themselves...(Chaimowicz 1985: 144–5)

CHAPTER 3

Gender Segregation and the Japanese Labor Market: Equal Employment Opportunity Law

In the preceding chapter on fieldwork conditions and methodology I detailed the range of social relations and multiple avenues of participation that were open to me through which I observed the intricacies of corporate life. Further, I pointed to how the intersubjective nature of anthropological knowledge might enrich our understanding of workers' lives in light of the contemporary transformations shaping Japanese society and corporate culture. Without losing sight of JCars, this chapter situates this ethnography in the wider historical, socio-cultural, legal, and economic climate of the late 1990s. Gaining an understanding of these transformations will enable us to analyze the community via the themes of gender, status politics, and ideology in finer detail in the following chapters.

This chapter explores the following issues: the changing image of salary men and office ladies since their emergence in the 1930s and 1950s respectively; the identification of patterns of gender segregation embedded in the structure of Japanese labor markets from the early 1900s to the late 1990s; the response of the 1970s management reforms to workplace gender issues; the socio-cultural factors contributing to women's exclusion in workplaces; the career tracking system introduced by the 1986 Equal Employment Opportunity Law (EEOL); the concerns of the 1997 EEOL over work-life balance and tougher calls on sexual harassment; and the 1990s economic downturn and its effects on employment practices within firms. In the last section of this chapter, I engage closely with the question of women's status in the workplace and begin to challenge the binary analytical framework through which anthropologists have traditionally understood women's status in relation to men's: inside versus outside; home versus work; informal versus formal means of power. This allows us to develop the analysis of status and occupational role in chapter 4 through an experiential and processual approach based on the detailed study of JCars' employment practices. Let us begin, then, by placing our understanding of salary men and office ladies in historical context.

Office Workers

The anthropological and sociological literature of the 1970s and 1980s on male white-collar workers focused on the relation between the nation-state, economic development, the company and the salary man (Dore 1987; Vogel 1979), and the structure of organizations and Japanese culture (Abegglen 1958; Clark 1979; Cole 1971, 1979; Dore 1973; Kinzley 1991; Nakane 1970; Pascal and Athos 1981; Rohlen 1974; Vogel ed. 1975). In this genre of writing, the Japanese company is seen as a symbol of the successes of the Japanese developmental-state model. However, this literature has taken a more critical tone on companyism over time, particularly following the challenge to Japan's economic success trajectory in the 1990s. Between the late 1980s and 1990s, Japan's role in globalization was explored via Japanese management relations with local staff at the firm's subsidiaries (Ben-Ari 1994; Hamada 1992; Ong 1987; Sedgwick 2001, 2007). Recently work has emerged on salary man masculinities (Allison 1994; Dasgupta 2000; Ishii-Kuntz 2003; Mathews 2003).

Feminist interest in the study of gender in organizations led to a proliferation of literature on Japan, including ethnographic studies (Carter and Dilatush 1976; Iwai 1993; Kelsky 2001; Kondo 1990a; Lebra 1981, 1984; Lo 1990; Matsunaga 2000; McLendon 1983; McVeigh 1997; Ogasawara 1998; Pharr 1990; Takenobu 1994). Labor market inequality and the perpetuation of gender segregation and status incongruity were explained in detail (Bernstein 1991; Bingham and Gross 1987; Brinton 1993; Cook and Hayashi 1980; Gordon 1985; Hunter 1993; Iwao 1993; Osawa Machiko 1994; Saso 1990; Sievers 1983; Skov and Moeran 1995; Smith 1987; Tsurumi 1984; Uno 1993); together with analysis of equality and legislation (Lam 1992; Mackie 1995; Osawa Mari 2000); the analysis of social policies formulated by a patriarchal, corporate-centered society and its implications (Morinaga 1995; Osawa Mari 1994, 1996; Ueno 1995); and feminization of employment (Gottfried 2003; Gottfried and Hayashi-Kato 1998; Osawa Mari 1993). In a global context, Western women also encounter gendering as a process of differentiation in workplaces (Acker 1990; Cockburn 1991; Kanter 1977; Pierce 1995; Pringle 1988; McDowell 1997).

Salary Man and Office Lady

This section briefly outlines some main points about white-collar workers as a sociological category. For a comprehensive social history of

Japanese clerical workers, I refer the reader to Konno Minako's (2000) *The Construction of OLs: Gender as a World of Meaning.* The salary man[1] (indicating a guaranteed income in the form of a salary) emerged in the 1930s, and became clearly identifiable in postwar Japan (Vogel 1963). This period in Japan's social history is associated with structural changes in Japanese society mainly in the economy at a time of industrialization and urbanization that gave rise to the nuclear family and a large new middle class. Salary men's lifestyles—daily commutes; transfers to subsidiaries; and the individual worker's ability to fully support a non-working wife—are said to resemble the lifestyle of *samurai* warriors of the pre-Meiji (1868) class system (ibid.; Ueno 1987a: S79). However, the *samurai* and salary man are seen to possess different qualities; whereas the *samurai* ideology upholds qualities of courage, boldness, and capacity for individual action (particularly in popular mythologies and media depictions), conversely, the salary man is bound to the organization, unable to make independent choices (Gill 2003; Mathews 2003; Plath 1964: 35; Vogel 1963). Nevertheless, the modern salary man carries the image of valor, as an efficient working machine, working hard for the household and nation's economy; the term "salary man," in essence, reflects a certain patriarchal cultural ideal of masculinity at the levels of the nation-state and corporate discourse and by individuals who uphold these moral values and work ethic (Morinaga 1995; Ueno 1995). However, this archetype has increasingly come under challenge in popular depictions, whether in fiction or in film (Dasgupta 2000).

We might also note that while the salary man ideology has supported the national discourse—the myth of Japanese uniqueness (Yoshino 1989)—the parallels between *samurai* and salary man also had purchase for Orientalist discourse in the West. As critics of Western Orientalism (Minear 1980; Miyoshi 1991, 1993; Miyoshi and Harootunian eds. 1989) have observed, the crux of Western discourse is defined by militaristic connotations (Dower 1986, 1993) and feminizing of Japan (Kondo 1990b). In Western media during the period of the bubble economy, parallels were drawn between Japan's aggressive wartime activities, and the corporate infiltration of Western markets to near saturation while Japan pursued isolationist economic strategies.[2] Particularly in light of powerful economic evidence that refuted the notion of an adversarial trade strategy, this commentary on Japan, fuelled by journalistic accounts, the work of revisionist specialists such as Fallows, Van Wolferen, Prestowitz, and Chalmers Johnson, and by political pressures in Washington DC, had parallels with the style of Orientalism directed against other non-Western states (Lowe 1991; Said 1995).

In the post-Restoration Meiji period (1868), as the counterpart to salary men, the Bank of Japan set the precedent in the recruitment of women as clerical workers (Encyclopædia of Popular Culture 1991: 92, hereafter EPC).

After the First World War, the trend intensified, and by 1920, 170,000 women were estimated to hold down clerical jobs. This number was equivalent to 10 percent of male salaried workers. The increase in the employment of women continued throughout the Second World War, when women were principally recruited as *hosa* (supplements) to men by the military.

Women became widely accepted as salaried workers in the 1950s, as economic development prompted the creation of numerous jobs for women in the expanding service sector that offered improved working conditions (Skov and Moeran 1995: 16). During this period such women were referred to as *shokugyō-fujin* (employed women) or as *ofisu-garu* (office girl). This shift within occupational structure, where previously women were confined to blue-collar jobs and only a minority of women occupied white-collar jobs, also reflected changes in industrial structure and the rise in educational levels of women (Lam 1992: 11–13). The contemporary terminology OL (*ō eru*) appeared during the early 1960s, owing much of its creation to the expounding of editors and writers for female weekly magazines (EPC 1991: 92).

In 2000, 29.3 percent of women (and 12.2 percent of men) were employed in clerical and related occupations, meanwhile approximately 19.2 percent of the whole workforce (including both sexes) made up the ratio of workers in clerical and related occupations (Ministry of Internal Affairs and Communications 2000). No significant change to this ratio has been indicated by figures for 2008: 29.2 percent of female workers (and 13.3 percent of men) were employed in clerical and related occupations, meanwhile 19.9 percent of the whole workforce (including both sexes) made up the ratio of workers in clerical and related occupations (Ministry of Internal Affairs and Communications 2008).

Gender Segregation in the Japanese Labor Market

A brief historical exegesis of differences in wage, employment status, and occupational roles between men and women is necessary to understand the significance of recent economic changes and their impact on the experience of social relations in the workplace. Japanese women throughout history have exhibited a high work participation rate, and contemporary figures show these rates to fall between the high rates of participation recorded for North America and Scandinavian countries and the lower rates of the rest of Western Europe (Brinton 1993: 3; Skov and Moeran 1995: 14–15) (see Appendix 2). In 2004, female workforce participation in the United States was 59.2 percent; in the UK 55.7 percent; and, Japan 48.2 percent (Ministry of Internal Affairs and Communications 2005).[3]

The evolution of economic activity over the past century is crucial for understanding the current position of women in the labor market (Bernstein

1991: 5). In the early 1900s, the textile industry dominated the economy in terms of production. A great proportion of employees in this industry was female, they were employed based upon their female characteristics: dexterity and being easy to control (Sievers 1983). Women working in the textiles industry were from the rural poor, the overspill from the agricultural labor surplus (Hunter 1993: 2); yet their low status as workers is equally attributable to women's status in Japanese society, derived from the gendered hierarchy and the association of women with the domestic sphere and their feminine roles (Hendry 1993a; Kondo 1990a). Contemporary statistics mirror this trend. A breakdown of Japanese women's occupations by type of industry (*Sōrifu* 1999b) shows agriculture, hunting, and forestry (29 percent); wholesale and retail, restaurants and hotels (28 percent); manufacturing (19.7 percent); financing, insurance, real estate and business services (9.6 percent); community, social, and personal services (6 percent); construction (4.2 percent); and other* (3.5 percent*—including mining and quarrying; electricity, gas, and water supply; transport, storage, and communication; and non-classifiable industries.) To avoid projecting a homogeneous image of female workers in contemporary Japan, we should note that there are two types of blue-collar workers: urban and rural (Cole 1971). Uno (1993) refers to the occupational identity of contemporary urban lower class housewives in assembling manufacturing parts at home (*naishoku*); and Martinez (1993) writes about rural women working in a seaside village as abalone divers.

Therefore, Japanese women have been employed historically in industries characterized by high turnover rates and low wages (Tsurumi 1984), but this is not peculiar to Japan as shown by studies of class and gender under global capitalism and the new industrial division of labor among female factory workers in free trade zones across the world (Fernandez-Kelly and Nash eds. 1983; Ong 1987; Tinkler ed. 1990).

Prior to the onset of positive changes in working conditions for women around 1930, Japanese heavy industry was characterized by a dual structure of firm size and wage differentials (Cole 1971: 37; Kondo 1990a; Vogel 1963). This meant that firms were either large with over 1,000 employees, or small with under 10 workers; wages were correspondingly either high or low. In contrast, light industries were not included in this dual system stemming from firm size; thus, the wage structure in light industries did not incur substantial differentials as seen among firms in heavy industries (Brinton 1993: 118–19). More generally, the differences in working conditions also presented a dual structure such that worker's conditions were characterized either by

... [H]igh wages, good working conditions, employment stability, and job security, equity and due process in the administration of work rules, and chances for advancement [or by] low wages, poor working conditions, considerable variability in employment, harsh and often arbitrary discipline and little opportunity to advance... (Lam 1992: 32)

Women did not reap any benefits from these structural changes during the 1930s for the following reasons: first, as they were excluded from employment in heavy industry (Gordon 1985; Tsurumi 1984) and second, because female employment remained concentrated predominantly in light industries that offered low wages (Brinton 1993: 120).

Long-term employment policies such as the lifetime employment system (*shūshinkoyō-seido*), seniority wage and promotion systems (*nenkō-seido*), and on-the-job training, which is said to distinguish current Japanese company policy from others in the world (Clark 1979; Cole 1971; Dore 1973; Nakane 1970; Rohlen 1974; Vogel 1963), have roots in the dual structure of the economy created during the interwar years. Initially, only a minority of male workers benefited from this structure as the majority continued to work in conditions of part-time, self, or family employment, but with time the proportion of the male population taking advantage of these wage structures tended to increase (Brinton 1993). That said, lifetime employment even when most secure represents at best 25 percent of the labor force. Women, moreover, were excluded from this lifetime employment, training, and seniority system because they were not employed in the heavy industries. This then placed women in a disadvantaged position in labor markets (ibid.).

By the 1960s and 1970s, lifetime employment, seniority, and on-the-job training became increasingly common among a greater number of firms. These characteristics expressed the absence of interfirm competition. For the company, reducing the turnover of workers was its primary concern, and the principal way of managing this was to guarantee workers job security and wages, benefits and prospects, better than those available in external labor markets and firms (Lam 1992: 31–2). This was reinforced also by a shareholder culture that avoided a preoccupation on bottom-line profitability in preference for a long-term perspective that allowed firms to plan strategically. The companies limited the number of entry points into the company, mainly at the lower level jobs; and higher level jobs within the company were filled through promoting and transferring workers in the company (ibid.). Economic logic and the support of labor unions—reinforced by the enterprise union model and a pattern of corporatist collective bargaining—oriented companies to envisage this type of management system (Cole 1971; Dore 1973). Gender-biased social insurance policies and tax planning modelled on a household unit (a wife who is looked after by a male provider) rather than on individuals as units also supported the stability of the lifetime employment system (Osawa Mari 1994; 1996).

Management Reforms in the 1970s:
Utilizing Women's Power

The increasing demand for female white-collar workers in the 1970s brought inevitable changes to the way management policies dealt with the issue of gender. The growth of the service industry was a direct result of the changing industrial structure that set the trend for the current working conditions facing many women. During this time, gender equality in the workplace became an increasingly important social issue.

Between 1975 and 1985, the demand for female workers increased significantly (Lam 1992). Events such as the 1973 oil crisis and the subsequent slowing down of the economy forced expansion in industries that had until then contributed only marginally to Japan's economic performance. The subsequent expansion of the female labor force was noticeable in the service, wholesale, and retail industries: this was not only driven by the rise in commodity prices, but also by the maturation of Japan's industrial structure as companies moved up the value-added chain. Also, the creation of many white-collar jobs and knowledge-intensive occupations allowed women to take up jobs that had been unavailable to them in previous years. This structural adjustment enabled women to enter clerical, professional, and managerial positions as specialists in the fields of banking, finance, and retail distribution in the service sector, and within information technology (Lam 1992: 70–1; *Nikkei* 09/02/98).

In fact, the cultural expectation for Japanese middle class women to be housewives has led to women's positions in the economy as powerful consumers. (Although this was less time contingent since women as controllers and managers of the domestic purse strings have always had a decisive discriminatory role.) In turn, this prompted the creation of jobs in marketing and design sections of companies, and within retail distribution and finance, as well as in manufacturing sectors of electrical appliance makers, automobile, and office automation makers (Lam 1992: 72). These new opportunities for women brought on by structural reforms in the education sector were enhanced further by changes in the representation of women in the media (e.g., the appearance of politically influential politicians such as Doi Takako). Increasingly, women were represented as individuals and this meant that women's association with the domestic sphere was less prominent in this particular context of representation. Meanwhile women were also able to appropriate meanings of "empowerment and emancipation" in their new identity as consumers (Skov and Moeran 1995: 27–31).

Changes to the internal conditions within the company also led to the increased employment of women. The seniority system of wage and promotion was stable in times of high economic growth, as workers in line for promotion could always be accommodated through the creation of new positions at the top of the hierarchy. However, the positions at the top became crowded in parallel with demographic changes that contributed to an aging population and with sluggish economic growth it made sense to reduce the number of workers who were due promotions. Women who were exempt from the promotional system and lifetime employment were thus recruited heavily to fill the jobs at the bottom (Lam 1992: 72–3). Clearly, the recruitment of women was a way for companies to preserve the internal structure of the company.

Companies have attempted to adapt to this increase in female workers and keep costs down, creating as a result "career development programmes," "women's project teams," and "female group leader systems" with an objective to improve morale among female workers (Lam 1992: 76–8; NSS 07/08/98a; NSS 20/08/98a). These policies do not attempt to compensate for gender inequalities in the workplace; conversely, they reinforce sex role segregation. According to Lam (1992), the function of the female "leader" is to act as a bridge in communication between female workers and the male supervisor, and thus it is a secondary role and not a formally recognized status.

Women's Ratio of Participation in Various Professions, 1990–2005

Between 2000 and 2005, the ratio of female administrators and managers in Japan has remained constant at 10 percent (women's share of earned income in Japan is 35 percent), this ratio is low compared to 33 percent in the UK (women's share of earned income in the UK is 39 percent) (Sōrifu 1999c; Sōrifu 2006). The Cabinet Office's basic data on gender equality in Japan gives the following figures (Sōrifu 2006: 23). Among national public officers women accounted for 1.5 percent of high-ranking managerial posts in 2003. In professional and technical fields, such as researchers in natural science and medical doctors, women are still under-represented, but they have been increasing in number: between 1995 and 2000 researchers increased from 12.9 percent to 13.8 percent; while medical doctors increased from 14.2 percent to 15.5 percent. Women's participation in politics equally remains low but shows increases in numbers: in the House of Representatives female membership accounted for 2.7 percent in 1993, which increased to 9 percent in 2005; in the House of Councillors membership of women stayed constant between 1995 and 2004 at 13.5 percent and 13.6 percent; and, in local assemblies between 1995 and 2004 female membership increased from 4.3 percent to 8.1 percent. Women's participation in the judiciary shows similar trends: the number of women judges between 1999 and 2004 increased from 10.4 percent to 13.2 percent; the number of female public prosecutors increased between 2000 and 2005 from 6.1 percent to

9.5 percent; and, the number of women lawyers increased between 1999 and 2004 from 8.4 percent to 12.1 percent. Perhaps, women fare better in some international positions. In 2005, 59.6 percent of all Japanese staff at the United Nations Secretariat were women (*Sōrifu* 2006). Yet in the same year women accounted for 1.5 percent of all Japanese ambassadorships; the first female ambassador was appointed in 1980. However, the number of women on national advisory councils and committees increased between 1996 and 2005 from 16.1 percent to 30.9 percent.

The Status of Female Workers[4]

Exclusion from Employment in Large Companies

The internal labor market system in Japanese companies that favors men and a long-hours work culture are responsible for the exclusion of women from entry into these labor markets. This is partly why the majority of women find themselves in temporary and part-time jobs. In other words, Japanese women do not choose part-time labor out of their "free will" (Osawa Mari 1993: 67–8). Moreover, women who are working for large companies face discrimination because

> The nature of the rules and practices governing the Japanese internal labour market are based more on personal characteristics than job classifications. Sex and age are often used as criteria for defining the rights and obligations of the employees. Discrimination against women constitutes an important basis for the employment practices characterising Japanese companies. (Lam 1992: 28)

Women receive substantially lower wages than men after controlling for differences generated by age, education, length of service, industrial and occupational distribution, and other productivity-related factors (Lam 1992: 47). In the 1970s among the advanced economies Japan had the greatest wage differential between men and women; in the late 1970s this gap widened in Japan while in the West the gap narrowed or stabilized (Osawa Mari 1993: 49, citing Koike 1991). Women tend to be concentrated in low-paying jobs (a horizontal component of occupational segregation) as clerical workers, service workers, production process workers, or unskilled laborers (p. 59). Female workers in the professional and technical category tend to be exclusively concentrated in the health service and in teaching (p. 51). Their productivity levels are said to be lower than men's for the same job, and even at the lowest ladder of the hierarchy within companies women start on lower salaries (p. 48). Companies justify lower wages by allocating women tasks and responsibilities that are different from male employees. Conversely, women who are given equal job content and the same level of responsibility as their male peers are nevertheless given lower wages. The Japanese company policy that pursues flexibility and efficiency while rigidly maintaining the traditional structure of social relations is responsible for women's structurally disadvantageous position. The cultural norms associated with gender as given below also give companies considerable leeway to maintain discriminatory practices, but this appears to be changing, as we shall see from the remainder of this chapter.

The discrimination of women in the workplace can be traced back to the structure of Japanese educational and familial exchange (Brinton 1993: 96). The cultural factors contributing to gender stratification in the workplace stem from greater familial investment in the son's education from an earlier age through to university level, as the same parents might be content with the daughter finishing her education at junior college (McVeigh 1997). The difference between junior colleges and universities, in terms of performance in league tables, is based on aggregates of standardized test scores that shows that even the best junior colleges only do half as well as universities. Thus, judgment of the various qualities possessed by an individual is implicated straightforwardly and unfairly by the name of the institution attended. Since women are the only employees to attend junior colleges, they are accorded lower status than men who attended universities, although there are university graduates among these women. In this, women are generalized as a category despite the variation among them. On the other hand, greater numbers of women are accessing university education at a high level (Eades et al. eds. 2005).

In Japan, education is a major marker of difference between individuals that affects the individual's employment prospects (Cummings 1980; Dore 1976; Hendry 1986; Passin 1965; Rohlen 1983; Sugimoto 1997) and the likelihood of promotion (Nakane 1970). The names of universities are seen to be a clear and unbiased guideline on which to base assessments. This is because "[t]he education system is seen as a pre-selection mechanism for the labour market" (Clark 1979: 146) that has been the only source of occupational allocation (Dore 1973: 293), although this procedure is changing with the availability of internship schemes that enable employers to assess ability in real terms. Moreover, the educational status of an individual is seen to represent something about his or her general background, that is, class orientation—and this is further connected to the assumption of certain acquired moral virtues. In this way educational status, class background, and moral virtues are intimately linked. This is because only well off families could afford to send their children to crammer schools (*juku*) in the preparation of the difficult university entrance examinations (cf., Rohlen 1980). In turn, the economic standard of the household is seen to be a good indicator of the moral values considered important by the parents, and in most cases this is reflected in the personal views and values of the children. The educational and professional status of the parents directly influences the degrees to which their children are inculcated with values that are culturally desirable to the middle class: social conformity and demonstration of high standards of etiquette and language (Kataoka 1987, cited in Sugimoto 1997: 46).

Before their stint as office workers, women in junior colleges are socialized in preparation for their intended roles as mothers and wives (McVeigh 1997: 149). This cultural tendency for women (workers) to be associated with the domestic sphere is mirrored in anthropological descriptions of office ladies: assisting and supporting men, in positions of little responsibility involving domestic tasks at the office—traditionally making and serving tea, wiping desks, typing up documents, faxing and answering phones (Clark 1979; Lo 1990; Ogasawara 1998; Pharr 1990). Women were then expected to resign from their jobs upon marriage (Rohlen 1974: 78). Women's retirement from the workplace at marriage is indicated by the dip in the M-shaped employment curve (Ministry of Internal Affairs and Communications 2005): the first peak in labor force participation is seen in the 25–29 age group (74.9 percent), which dips in the 30–39 year group (63 percent), there is a steady rise until it peaks again in the 45–49 age group (73.9 percent). Consequently, women are given low status in the workplace, and become associated with the domestic sphere rather than with the workplace and their roles are limited to domestic tasks. However, the higher status of women as they become mothers and wives compensates for women's lack of status in the office (Iwao 1993; Lebra 1984; Roberts 1994).[5] A woman's status fluctuates with age; as working while married is inconsistent with middle-class values (Rohlen 1974: 78); and thereafter being a middle-class housewife can confer prestige, insofar as it is crafted to mirror the status of the professional working husband (Hendry 1993a: 225; Imamura 1987; Vogel 1978). The mother's role in the education of her child also accords recognition (White 1987), yet this can place considerable psychological pressure on mothers (Fujita 1989; Lock 1988).

Gender specific cultural norms concerning marriage also support gender stratification in Japanese companies. The regularity of the life course in Japan (Brinton 1992: 80) generates a strong consensus for women to marry at a certain age. Currently, as my subjects at the office told me, the social norm for a woman's appropriate age at marriage (*tekireiki*) has risen to between 26 and 28 years old. Twenty-four is seen as too early and twenty-nine as late. Correspondingly, macro statistics indicate that the average age at first marriage of the bride is 26.2 and 28.5 for the groom (*Sōrifu* 1998a). There is also a steep increase of women in their thirties and men in their late thirties and older who have never married (ibid.). This trend has occurred in part because women have put off marrying to extend their careers; it gets around mandatory retirement at marriage (Osawa

Machiko 1994: 3). Further, now there is greater variation shown in the age at which people marry for the first time than ever before (*Sōrifu* 1998a).

It is increasingly acceptable for women to marry later than say 10 years ago and there is evidence of this flexibility in the life course of individuals, but the notion of a highly regulated life course (Brinton 1992) still exists. Although women are not as a rule expected to resign when married, this strong cultural notion of the appropriate age to marry leads to a cyclical process whereby employers expect less commitment from female workers who are therefore expected to leave after marrying. Also women give pregnancy and motherhood as reasons for retirement in place of marriage as once was the case (Yamashita 1996). Hence, employers are reluctant to invest in training for women (Brinton 1992: 97–8). Again, women are excluded from promotional prospects that involve developing close working relations with mentors inside and outside of the workplace, on basis of the assumption of early retirement (pp. 102–3). In this system women who aspire to careers are automatically disadvantaged.

Equal Employment Opportunity Law (EEOL)

The high degree of gender stratification in the labor market as discussed above, based on women being valued as mothers and wives, with marriage seen as a dominant cultural ideal, is linked to women's low status in the workplace. Before looking at women's status in JCars, it is necessary to examine the judicial measures aimed to better women's status in the workplace. The changes to the company's recruitment process and structuring of social relations through the tracking system, envisioned by the EEOL which was enforced in 1986, is discussed in this section.

Under the EEOL, invariably and somewhat problematically, equality, which is a fundamental right of an individual, was construed in patriarchal terms as "welfare" extended to women by the state (Lam 1992: 101). This sits oddly alongside the 1947 constitution with its explicit guarantees of gender and employment equality of opportunity. The EEOL was instigated amidst the strong opposition from companies. The government was insistent in pursuing this legislation because the EEOL would be a means to contain the increasing number of claims of unequal treatment in the workplace made by women in the courts. These demands for equality were perceived as a threat to the social order because litigation as a response to an undesirable situation is considered disreputable by Japanese cultural values. Numerous malpractice suits have been brought against employers under the following conditions of discrimination: in equal pay and fringe benefits, retirement policies, transfers, maternity, and menstrual leave (Cook and Hayashi 1980; cf., West 2005). These claims testify to the growth of women's self-consciousness and determination to be treated equally in the workplace (p. 3).

Two forms of sanctions govern the practice of the EEOL (1986): prohibition (*kinshi-kitei*) and exhortation (*doryoku-gimu-kitei*). The EEOL prohibited employers from engaging in unlawful discriminatory practice in the following areas of women's employment: basic training, fringe benefits, retirement age, resignation, and dismissal. Changes in these areas of management practice seldom incur repercussions for the core of the management system. At the same time, revisions in these areas of practice do not completely relieve women from their conditions of discrimination (Lam 1992: 104). The law is deliberately vague in indicating the exact level of equality to be achieved in the workplace. Moreover, companies were merely urged to provide equality or discouraged from discriminating in the following areas: recruitment, hiring, job assignment, and promotion (Lam 1992: 101). The EEOL therefore protects employers against legal action in the areas that might be of most concern to women.

The cautionary guidelines administered by the EEOL have induced some policy changes in the management policy of companies. Changes have been most evident in the areas of recruitment, conditions of employment, and retirement, whereas assignment, job rotation, and promotion have received the least vigorous revision. For instance, in the area of recruitment, the number of job placements among women has increased during 1985–1989 relative to men. Also, more women have been acquiring jobs in large companies where they had previously been excluded. The amount of women working for the same company for over 10 years has increased by 17 percent in the years between 1980 and 1990 (Iwao 1993: 161).

Introduction of the Career Tracking System, EEOL (1986)

The introduction of the new career tracking system (*kōsu-betsu-koyō-seido*) in large companies divided new recruits into clerical (*ippan-shoku*) and managerial tracks (*sōgō-shoku*). The clerical track is characterized by less complex and more manual jobs, lower pay, fewer job rotations and only limited transfers, and promotion limited to lower level or local management positions. In contrast, jobs in the career or managerial track are seen to require complex judgments such as business negotiations, personnel management, designing or developing products, and planning company policies or strategies. Further, there is no limit to promotion and the individual is subject to job rotations and transfers (Lam 1992: 128–9).

This career tracking system makes claims to sexual equality as, supposedly, merit is used to differentiate recruits into separate tracks. In reality, tracking is based on the willingness to accept job transfers to other subsidiaries and this is something women are reluctant to do (Lam 1992: 131). Moreover, "men seem to be assigned to the career track automatically whereas women are selected for it exceptionally" (ibid.). The tracking system clearly is an alternative division based on sex: survey results from 1987 based on career tracking in 40 companies found 1.3 percent of women and 99.0 percent of men on the career track (ibid.).

For women the EEOL changed the way they are recruited and this introduced official status differences among women, while it did not necessarily put women on an equal footing to their male counterparts in the company. In effect, the tracking system complicated the relationships between women. Although these status distinctions between clerical and career track positions were official, the company used arbitrary guidelines in the evaluation of work performance. Educational background remained the principal structural barrier to promotion for women with fewer years of education, even though they might have outperformed women with higher educational qualifications, evaluated in terms of length of service and amount of responsibility handled in the job.[6] This further complicated social relations between women already compounded by factors of age, tenure, and status.

The women in the career track also face complications in their social relations with male colleagues in the workplace. Women in the workplace are laden with traditional gender role expectations that confine them to an allocation within supporting job roles, that is, when women take up positions in the career track, these traditional gender role expectations were maintained within the perceptions of male colleagues. Takenobu (1994) notes an incident of discrimination related by a woman at a new recruit's training session. First, a male colleague of the same year of entry exclaimed surprise and indignation when he found a woman was on the same pay scale as him (p. 44). Second, her assertiveness shown at the canteen in refusing to serve tea and food to the male colleagues of her year of entry resulted in a culmination of bad feelings between them: she found herself alienated from the group. Alienation from colleagues in the same year group could be troubling for women in terms of career prospects, because the horizontal level of organization is thought to provide an invaluable basis for information exchange about other workers and departments.

Following 10 years of the EEOL (1986), a survey (1995) completed by approximately 700 employees of private companies with Enterprise Labour Union Membership outlines the general working condition in large firms (*Rōseijihō* 13/12/96). Delineating both men and women's perspectives, the survey deals with six main areas: (1) recruitment and hiring; (2) retirement; (3) wage differentials; (4) ban on overtime and night-shifts; (5) differential treatment of women and men on the job; and (6) the two-tier tracking system. The source of this survey is a business management journal read by human resources managers of large corporations. While the representativeness of this survey can be questioned on the basis of sample size, I include this general overview of the perceptions of gender in/equality in an attempt to relate the experience of my subjects at JCars to a wider societal level. This survey effectively brought to the fore the following issues regarding the 1986 EEOL:

(1) The guidelines to curb sex discrimination at the stage of recruitment and hiring were not followed rigorously by companies, and once employed the majority of women expressed experiencing degrees of differential treatment, notably in the assignment of tasks, in lower expectations of male managers, as well as in lower wages.

(2) With regard to retirement at marriage, women overall did not experience pressures to resign, but, equally, women expressed uncertainty as to how to read this pressure, since it takes implicit forms.

(3) Women strongly felt that wage differentials exist between men and women; a greater proportion of men thought that gender does not determine wage distinctions. Women thought their lower wage was directly connected to their low-level tasks allocated on account of gender, these did not merit high work evaluations. Women also thought they received unfair treatment at work appraisals while experiencing fewer opportunities for promotion in comparison to men. Women are excluded from family and housing allowance as part of their wage package, although men received these automatically. Over half of the men surveyed thought gender-based distinctions were irrelevant to wage differentials, but were rather attributable to job type and content.

(4) Two ideas came across regarding the regulation of working hours: First, men's working hours should be regulated in accordance to women's. Second, women's work hours should only be regulated during pregnancy and childbirth, and for the remainder of the time, women should be allowed to work equivalent hours to men.

(5) A majority of women experienced differential treatment on an everyday level, giving rise to feelings of alienation. They were excluded from access to necessary information, participation in meetings and business trips. Moreover, women thought others played down their roles, particularly when they were given greater levels of responsibility. The male view toward women's feelings of exclusion, alienation, and discrimination demonstrated a relatively simplistic understanding; attributing tea pouring duties as the main cause of women's negative experience in the workplace.

(6) Half of this survey's respondents did not have a career tracking system implemented at their place of work. Where the tracking system was in place, as a result, 60 percent felt that women's job content was made equivalent to men's. However, they expressed concern over the attitudes of their bosses (who did not take women seriously and did not spend enough time in their explanations about tasks, etc.), as women thought this interfered with work performance

and morale. The career tracking system, nevertheless, seemed to extend the length of women's careers (from an average of 3–5 years to 8.3 years).

EEOL Enacted (1997), Enforced (1999)

The strengths and weaknesses of the 1986 EEOL inferred by the survey result above coincide with the experience of my subjects. The persistence of gender inequalities generated a new version of the EEOL that was enforced toward the end of my fieldwork. The guidelines were enacted in June 1997, and enforced in April 1999. The following social trends facilitated the amendment (MoL 1998: 1): 女性の雇用者数の大幅な増加 (profound increases in the total number of female recruits); 勤続年数の伸長 (the prolonged length of service [of women] in the workplace); 職域の拡大 (the expansion in the type of jobs [for which women applied]); 女性の職業に関する国民一般の意識 [変化] (the changing awareness among the general public regarding women's work); 企業の取り組み [の変化] (the changes to the way companies tackle [policy changes with regard to the equal treatment of female workers]).

Although unmentioned in the MoL's document above, the concern over falling birth rates and its connection to the revisions should be mentioned (*Sōrifu* 1998b):

> 1996 年の合計特殊出産率 . . . は現在の人口を将来も維持するのに必要な２. ０８を大きく下回る１. ４３となっている. . . .　　出産率の低下は女性の晩婚化の進行と生涯未婚率の上昇によるものであり、その要因として、特に女性の社会進出が進み育児の負担感、仕事との両立の負担感が増加している。

> [The total birth rate has fallen well below levels needed to maintain the current size of the population (from 2.08 to 1.43 [1996]) into the future, due to a progressive trend for later marriages among women and a rise in individuals opting out of marriage altogether. The main cause of the above is directly related to the escalation of women's participation in the workplace, which presents women with a greater burden in childcare, making it difficult to balance childcare with work.]

Tomiyama-*shunin*, a personnel manager in his early thirties at JCars, was in charge of monitoring the requirements of the 1997 EEOL in relation to company policy. In a response to my interview, he defined the main area of difference between the first and second versions of the law: he predicted the new method of enforcement would have some impact to ensure tighter policy revisions in companies. In the 1986 EEOL, areas of the law that were softly enforced through "exhortation" (recruitment, hiring, assignment, and promotion) were revised in the 1997 EEOL, to the status of "prohibition." Examples of *prohibited* forms of discrimination against female workers indicated in the guidelines include (MoL 1998: 4–6):

Recruitment and Hiring

(1) Recruiting only male applicants;
(2) Using gendered terms in recruitment advertisements such as "salesman" or "waiter";
(3) Advertising themselves as an equal opportunities employer, while only accepting applications from male candidates;
(4) Advertising the exact numbers they are looking to recruit;
(5) From finalizing the recruitment of men, and then recruiting women at a later stage to fill the remaining positions;
(6) Fixing the rule that if an age limit for the job is listed, the women's maximum age cannot be lower than that advertised for the men's;
(7) If the position specifies unmarried applicants, this cannot be restricted to women;
(8) Companies expecting the employee to commute from their parents' residence cannot confine their expectation to women;
(9) Age discrimination against women who have spent extra years preparing for their university entrance examination or taken a gap year;
(10) Mailing company information booklets exclusively to male candidates;
(11) Prioritizing the mailing of company information booklets initially to male candidates and then later to female candidates;
(12) Holding separate open days for male and female candidates, or from holding open days exclusively for male candidates;
(13) Holding separate entrance examinations exclusively for female candidates;
(14) Issuing extra examinations to female candidates.

Assignment and Promotion

(1) Denying women equal chances of promotion;
(2) Restricting women from further promotion;
(3) Withholding promotion on the basis of marital status;
(4) Offering promotion only to a certain level, for example, on the basis that they have children;
(5) Both women and men should be promoted at the same time (e.g., all employees must be considered for promotion after X number of years).

Additional issues dealt with in the 1997 EEOL are the prevention of sexual harassment in the workplace; the abolition of restrictions on overtime and late night work by women; and restricting late night work by workers engaged in child care or family care (*Sōrifu* 1999). Dealing with sexual harassment is now a legal requirement. Therefore, if it arises, its resolution will no longer be left to the discretion of upper managers within the company (*Nikkei* 14/06/97). However, a clear definition of sexual harassment is still necessary, and this judicial policy seems to be an empty formality, stated only to raise awareness of sexual harassment in Japanese workplaces in alignment with Western standards, as Western and Japanese companies interact in the global economy (ibid.). The previous ban (EEOL 1986) on overtime for female workers based on a paternalistic protective concern for their welfare was a barrier to the promotional prospects of women. Due to this ban, women were not given challenging tasks. Lifting this ban on overtime is considered a means of achieving equality for women in the workplace.[7]

Furthermore, the government offers 調停制度の改善 "an improved consultation and arbitration service" to enterprise owners who formulate active corporate measures to relieve existing disparities in the experience of the workplace between male and female employees. The government will provide for the voluntary in-house resolution of complaints: assisting dispute settlements through the Director of Prefectural Women's and Young Workers' Office; and offering relief from disputes through mediation by the Equal Opportunity Mediation Commission. The MoL will also provide the necessary administrative guidance to ensure a strict enforcement of the EEOL in the form of advice, instructions, and recommendations (*Sōrifu* 1999a). As a means of reinforcing this new standard, the law aims to regulate companies exercising discriminatory practices through threat of public exposure (*kigyōmei-kōhyō-seido*). The will to preserve a good public image and pressures to keep pace with other companies should function to regulate the practice of companies (Fox 1996: 2). But the extent of the success of this measure is met with doubt given the influence of large companies on political matters (*Nikkei* 14/06/97).[8]

The Company, Society, and EEOL

The EEOL cannot be enforced effectively in companies if society does not support working women. Companies must also provide women with an environment wherein they receive equal treatment to men as well as childcare facilities for career women. Furthermore, in addition to women, it is necessary to focus on the working conditions of men if gender equality is to be achieved (Ueno 1995: 216). Similarly, Tomiyama said: 働く女性を助けるには男性の働き方を今より楽にしなければならない、それには日本社会全体の考え方を変えないといけない (To help women, we must relax the way men work, and for this to happen it is necessary to change Japanese social norms). Indeed, at an EEOL seminar for companies given by the MoL's Women's Employment Bureau (*rōdōshō-josei-kyoku*) that Tomiyama attended, a female section leader who was an official speaker said: EEOL なんか関係ないですよ、夫が世間を気にせずにご飯作りに帰ります等と言えないかぎり．．．それが言えたら女性も平等な地位が貰えます (The EEOL is irrelevant [i.e., the revision cannot change the status of women in the workplace] unless a husband can say without feeling uneasy about *seken*[9] that he must leave the office to prepare supper, etc.).

This notion that men need to balance home life with work life was raised at JCars. A salary man is typically expected to make a full commitment to working for the company by sacrificing his life at home (Allison 1994; Lebra 1981; Ogasawara 1998; Rohlen 1974; Vogel 1963). The salary man, called a "*kaisha ningen*" (company person) should be seen acting out their appropriate role, thus men at Tozai Bank obscured their private roles as husbands and fathers (Ogasawara1998: 87–8). In this reference, the term *kaisha ningen* reflects a positive and desirable image for men. But during my fieldwork managers in their thirties described themselves as *kaisha ningen* not so much as to demonstrate their commitment to work as to emphasize a feeling that they were office androids controlled by the organization unable to give priority to their wives/family/children. During my interviews with managers in their thirties, they volunteered information about their private family lives and asked about my own family situation (presumably because they knew [of] my father). Many managers displayed photographs of their family on their desks, framed or placed underneath the clear plastic covering on top of work desks, or on their desktop and screen savers on their laptop computers. Their self-descriptions consisted of an image of a man of principle, committed to hard work, but equally to his family (Ishii-Kuntz 2003), which they called *mai hōmu shugi* or "my home-ism" (contemporary salary man masculinity is discussed in chapters 4 and 6).

Yuasa-*shunin* in the personnel department would not work longer than necessary, that is, stay at work unnecessarily in order to give moral support to others in his department doing overtime. In addition, he would not go out drinking with his friends or colleagues more than once or twice a week. He joked that he was afraid of incurring his wife's wrath; actually, he enjoyed getting home to play with his young daughter before her bedtime. Yuasa thought his type of worker, someone committed to the notion of *mai hōmu shugi*, was currently common among salary men. Among Yuasa's generation of workers, then, the term salary man had (in a loose sense) lost its former hard image of workaholics, which was an ideal to the generation of managers in their forties and fifties. The image of the salary man seems to be reverting to the image of workers who balance work and family time (see Vogel 1963).

Taniguchi-*kachō*, a manager in his early forties, thus older than Yuasa, admired by women in the personnel department for his humanitarian and *mai hōmu shugi* principles, worked the longest hours in his section. At times when I stayed behind at the office until 7 and 8PM to photocopy relevant literature (the machines were free at this time so I was not getting in people's way), before leaving, I would occasionally chat casually to Taniguchi. When I mentioned the late hour, he would grimace and retort that he was living the life of a salary man, and at times he would use the term *kaisha ningen* in the ironic sense to describe his commitment to work. Describing salary men as *kaisha ningen* maintains its cultural relevance, but now it seems that the ideal of *mai hōmu shugi* is stronger.[10]

Further evidence of women's status in the workplace becoming conceptualized differently to how it was in the 1980s when the EEOL was first issued concerns the response of the courts to female plaintiffs' claims. Since Cook and Hayashi's (1980) accounts of difficulties encountered by women who presented their cases of unequal treatment in the workplace at employment tribunals during the 1970s, the judicial verdict of the 1990s has shown increasing sympathy toward female plaintiffs (*Nikkei* 19/07/97). This signifies a greater recognition at judicial level for equivalence in the workplace. For example, 28 part-time female staff in Nagano Prefecture demanded equal pay in line with levels received by permanent staff, on grounds of being long-term employees. In March 1996, the plaintiffs won and received 80 percent of the wage given to permanent staff (ibid.). In another case, 13 female workers accused their company of discrimination in terms of promotion, as their male colleagues of the same year of entry were promoted to section head while they remained stationary as clerks. In the ruling, the court ordered the company to give these women section head status in pay, if not in actual title. This case was seen as a betterment of previous verdicts in which companies were only ordered to reimburse the plaintiff for the cost of the court case (ibid.). In a sexual harassment case tried in 1997, the court ruling accepted the delayed reporting of the incident, on the grounds that the notion of sexual harassment can be interpreted widely, hence the leniency reflects the court accounting for differences among behavior of women exhibited after the assault (ibid.).

The combined effect of stricter EEOL policy, changing cultural views of the role of men at work and at home, and changes in the judicial treatment of inequality experienced by women in the workplace as given above will be examined in the context of recent economic changes in the following section, in order to examine how economic restructuring impacts on the structuring of social relations in the company.

Economic Downturn and Restructuring in the 1990s

Social and Structural Effects

During the 1980s and 1990s, Japanese companies faced two notable problems: First, the downturn in the economy that forced a strict reassessment of labor costs. Second, the aging profile of their personnel (Clark 1979:

248). Personnel policy has been the main point of focus for revision in companies whereby the lifetime employment system, and seniority wage and promotion systems, are being abolished or altered toward recognition of qualification and competence (Holzhausen 2000). The rate at which men left the workplace was 14.9 percent in 1997, and the trend is set to increase (this figure is based on firms with over five employees) (*Nikkei* 07/08/98a).

The unemployment rate is high, as is the rate of unemployment accounted for by redundancies, which is a direct result of companies restructuring through job cuts (*Nikkei* 01/08/98).[11] Under the lifetime employment system, companies are bound by non-contractual agreement to provide jobs for life to their employees. Therefore companies have responded by creating redundancies either by encouraging older employees to retire early (*NSS* 03/08/98) or by asking volunteers to take early retirement or temporary leave (*NSS* 12/08/98). The greatest job losses are reported among middle-aged male workers (*Nikkei* 07/08/98a), and suicide rates and divorce rates are increasing.[12]

Since having been made redundant, one middle-aged manager lived anonymously in a homeless shelter as feelings of shame prevented him from returning home to his wife and daughter (BBC 2, Bubble Trouble 2: 09/01/00). To aid this growing body of retrenched salary men, the government proposed to extend financial support to new small and medium sized ventures under the likelihood that these will soak up these individuals. The recipients also include housewives, students, and OAPs (*NSS* 16/09/98).

The changing economy also has implications for new graduates. As companies cut back on recruitment, only 65.6 percent of newly graduated job seekers (out of the total 529,000 university graduates) were able to find employment within the same year (*Nikkei* 07/08/98b), as opposed to 81.3 percent in 1991, during the tail-end of the economic bubble. Female graduates with higher education have suffered the most (*NSS* 20/08/98b). For example, a comparison (1998) of the projected and actual recruitment levels of the nine most successful Japanese trading companies shows that the number of career track workers hired undermined the target by 20 percent (*NSS* 20/08/98b). In this same projection, although there was no plan to slice the proportion of female recruits taking up career track positions, female career track recruits accounted for only 6 percent of total positions filled by new career track workers (ibid.).

As a response to the drop in levels of intake of new recruits, internship programs are being set up in large companies. Large traditional companies in western Japan, mainly in Osaka, Kyoto, and Hyogo Prefectures, have recently introduced internship schemes giving university students work experience during the summer holidays (*Nikkei* 15/08/98). Internship opportunities as they exist in Western companies help to develop creativity and inspiration while creating a smoother flow in the transition between university and the workplace. It is believed in future that not only large but also small Japanese companies will incorporate the internship scheme within existing recruitment methods (*Nikkei* 06/08/98).

Restructuring in the Manufacturing Sector

So far, voluntary early retirement has been the main way through which companies created redundancies at the top of the organization. Ashikawa-*jichō*, a section head of the overseas department, brought several newspaper articles to my attention that focused on case studies of changing personnel policies. Ashikawa said these examples illuminate the economic situation effectively and they situate JCars in a wider sectoral context. Company responses to the economy through the restructuring of personnel policy can be seen in the following example of Hino Jidōsha (a truck-manufacturing firm based in Hino City, Tokyo Prefecture) (see *NSS* 12/08/98).[13]

Hino has a total of over 9,000 staff; just over half this number work in the *kansetsu-bumon* (departments usually not directly effected by economic fluctuations). Following a 47 percent decline in production levels, Hino took dramatic measures to cut personnel in the *kansetsu-bumon*: a total of 150–200 personnel were offered one year's leave (*kikyū*). This policy operates on a voluntary basis and during this leave workers receive 80 percent of their usual wage, and an additional ¥50,000 per month in compensation, to be used for the workers' "education": attaining qualifications by attending vocational courses (*shikaku shutoku*). This method of compensation for temporary job losses functions to benefit both individuals and the company, as the worker returns to the company after one year having gained new or improved existing skills. At JCars this kind of restructuring did not take place, but some workers, particularly male managers, expressed concern over the possibility of similar changes occurring to them.

A white paper (*rōdō-hakusho*) places the above example in context of an aggregate analysis of the economy (see *Nikkei* 07/07/98). In brief, this study states that the first stage in the adjustment of firms to the economic crisis was through decreasing working hours, but by 1996 further change was introduced by stricter measures including job cuts. Also noted was the decline in the performance of the manufacturing sector in contrast to the expansion and increasing dominance of the service sector as well as the aging of the population. The paper proposed further measures for dealing with the economic situation: to increase the employment mobility of workers between different sectors of the economy; and to make use of the residual non-working population, that is, retired individuals and women.

Amidst economic and cultural change, internship schemes and the greater acceptability of mid-career changes reflect an emerging pattern of flexibility in corporate attitudes toward the selection of workers (see Beck and Beck 1994). This change may be read in two ways: First, as evidence of an egalitarian merit and experience-based evaluation system. With the abolition of the lifetime employment system, there will be a change to a system of promotion whereby individuals will be assessed on a meritocratic basis rather than on seniority or status.[14] Some large firms, such as Sony, Toyota, and Honda (because they are in the public eye), have installed these changes. Similarly, at a trading firm, the age at which men can qualify to attend the management training course was lowered from 40–45 to 30–35 (*NKKS* 17/09/98). However, smaller or lesser known firms are unlikely to encounter public scrutiny (said my subjects) and will be slower to change.

Second, these changes can be read as a response to the recession. Limited budgets and smaller intake of recruits force companies to select only the best workers. Thus, by recruiting interns on a trial basis the company accrues cost-saving advantages. The hiring of mid-career workers also enables firms to respond swiftly to changes in business conditions. For this reason mid-career workers who can contribute immediately to the business are hired as a corporate strategy to weather the recession (Ministry of Health Labour and Welfare 2002). The hiring of mid-career workers is expected to continue in the future. Recruitment methods are diversifying to include year-round recruitment, hiring by job type, trial employment, and hiring for a designated place of work. For workers, the rate of job changes among the young reflects a diversification in workers' attitudes. In the culture of the workplace, in terms of the turnover of male employees, the greater acceptability of mid-career changes can be interpreted as a feminization of the workplace as men's experiences are starting to echo a trend in Japanese women's careers.

Implications of Restructuring

Ethnographies of Japanese companies often focus on the harmony characterizing the organization (Abegglen 1958; Clark 1979; Cole 1971; Dore 1973; Iwao 1993; Lo 1990; Rohlen 1974) that is achieved through the "individual's total emotional participation in the group" (Nakane 1970: 20). Harmony in the workplace is thus construed as being a unique characteristic of Japanese companies, although U.S. firms such as Hewlett-Packard and Korean firms have modelled their organization on Japanese practices. High social status is also ascribed to workers in large organizations, reflecting the high barriers of entry into large firms. The cultural ideals of harmony and status can be interpreted as being superimposed upon an economic logic that oriented management strategy in the period between the 1960s and early 1990s. The harmony characteristic of Japanese companies is in itself not an attribute that is derived solely from cultural values of Japanese society as it is often explained; rather it has a strong economic explanation to support it. Currently, however, this stable economic model of Japan is fraying at the edges, offsetting the changes in personnel management policies in companies as shown above. Thus the model of the Japanese company as a harmonious organization is all the more in need of examination (Osawa Mari 1994, 1996; Tabata 1998). In this light, we will re-examine the changes to women's status in the workplace.

Women's Status Reconsidered

Women are to the inside (domestic sphere) as men are to the outside (public or work sphere). This view has been central to studies of the workplace. Office ladies, and more generally, female workers in the Japanese labor market have been conceptualized as oppressed victims of patriarchal society (Bingham and Gross 1987; Brinton 1993). Office ladies describe young and unmarried; married with children; and older, single, or widowed women (Lebra 1976). Collectively they were seen as a source of cheap labor, receiving low pay compared to male office workers with no access to higher positions (ibid.), and they are limited to occupational roles that emphasize domestic tasks (Lo 1990; Ogasawara 1998; Pharr 1990).

Dorrine Kondo (1990a) looks at middle-aged, female, part-time workers in a sweet factory in Tokyo and their methods of negotiating power within informal structures of the workplace. Here women gained power by assuming the parent/mother role to the child's role taken on by younger men. According to formal structures, women are peripheral to men, but by contrasting formal and informal structures, Kondo shows how women are not subordinate all the time. Yuko Ogasawara (1998) makes the same distinction between formal and informal structures in her study of female office workers in a large Tokyo bank. Formally, women are accorded low status based on employment status, wages, and occupational roles, but by drawing on the binary that women are associated with the outside and men with the inside, Ogasawara shows how women exert power and influence through informal activities such as gossiping despite their peripheral position.

Takie Sugiyama Lebra (1981, 1984) does not distinguish between informal and formal structures, but maintains the inside/outside dichotomy of gendered spheres. Career women are described as outsiders in a world of men (also see Valentine 1990), but find compensatory opportunity or advantage by virtue of their outside status and role ambiguity, whereby women are less constrained by proper channels of communication than men (Lebra 1981, 1984). In a similar view, accepting the sharp delineation of male and female domains, women are described as relatively free from entanglement in the male worker's intricate network of social relations (*shigarami*) of the organization therefore having an advantage over corporate men (*Nikkei* 05/01/98).

In addition to role ambiguity, women's status in the white-collar workplace is defined by status ambiguity (Kitahara 1994). These positions seem to best explain women's status in large companies. Status ambiguity arises from male and female office workers in large Japanese companies constituting the economic elite (Ishida 1993; Lebra 1993; Vogel 1963). As a status group, their qualities are defined by achievement (not ascribed characteristics), power (not prestige), and class of wealth (as opposed to "status of honor") (Lebra 1993). Female workers who enter this status group find, nevertheless, that ascribed characteristics, mainly gender, when combined with socio-cultural norms expected of women's lives, produce status differences between men and women in the office. Thus, formal structural inequalities in wage and promotion give women low status in comparison to men, producing the effect of status ambiguity.

Despite having made the transition from domestic to work sphere, from inside to outside, women still maintain their status as outsiders, because, via cultural norms governed by patriarchal, bureaucratic ideals, men are seen to constitute the inside of the male-dominated organization. However, despite the automatic assumption of women's association with the home, the initial status of women as insiders in the home is actually more complex, because at the point of marriage, when entering into her husband's household, women are outsiders (Napier 1998). Women are accepted into and adjust to their status as insiders, through their roles as wives and mothers, as men are seen to be the true insiders in the patriarchal household (Martinez 1998: 7–8) and Kondo (1990a) nicely described this process as mentioned above; but JCars has no real process for this. Rather, what we observe at JCars is a different process of inclusion. To understand this process it is vital that we deconstruct the perspectives that construe women as outsiders. Is this how men predominantly perceive the workplace? if so, to what extent do female workers accept this view? or is it women who fundamentally assume this distinction through sustained gender beliefs that translate into behaviour (Ridgeway and Correll 2004)? Moreover, have wage and promotional structure, occupational roles, and socio-cultural norms remained static?

The issue of women's experience of status ambiguity looms large at JCars. Men as well as women said that women are on an equal footing with men, that is, women are not always or no longer outside of men in the workplace, yet equally I heard the contradictory statement of discrimination against women, but this was a minority voice. Occupational role is crucial to the understanding of women's status in the workplace, that is, how workers define and interpret their roles, and talk about "work"—in terms of content and amount assigned. Once women's jobs were considered to be uncomplicated and supportive, but during my fieldwork I saw how women were given tasks of their own, equalling their male colleagues in complexity

and level of responsibility,[15] and these provided a means for women to negotiate their status vis-à-vis men and among themselves. What this means
is that women were experiencing wage differentials to men and obstacles
to promotion, yet they were experiencing changes in their occupational
roles that allowed them to enhance their status and influence in the firm.
Although this change in occupational role was not reflected in changes
to most women's official rank, it changed the way women conceptualized
and experienced their status. This also reflected the changes taking place
among some managers' views toward women.

This change observed within JCars is a reflection of wider social changes
to women's occupational roles and structural positions in terms of pay and
promotion: for example, in forward-looking companies, often in Western
multinationals with offices in Japan, women are being headhunted (*Nikkei*
07/01/98) and pursue enviable careers. At the office of a national newspaper
in Tokyo where I spent two weeks in April 1998, women were on par with
men in terms of assignments and pay, and they did not consider their status
any lower than men's nor did they describe themselves as outside of the
male career structure.

On the whole, women's status in the office will appear low if the analysis
focuses on macro level social indicators that emphasize structural points: for
example, high barriers to entry into competitive labor markets; low percentage
of women in managerial positions; difficulty in gaining access to training and
opportunities for career development; and skewed wage differentials between
men and women. These structural points used as objective markers of status
often echo male manager's perspectives who produce the dominant discourse
of the company and it is this group of men who maintain the insider's distinction of their own positions within the workplace, thus construing men as
insiders and women as outsiders. However, not all points of view encompassed
by this dominant discourse produce effects or influence decisions concerning
the wage and promotional structure of women, yet these points of view were
influential to the way women were allocated tasks, and helped some of them
to have long-term careers. What follows naturally is to consider the status of
women from the point of view of those whose voices do not occupy prominent
positions in defining the dominant discourse, but are equally important to
women's experience of the workplace. The task of understanding occupational
role—of how it is redefined and legitimized—in the light of recent legal and
social changes is crucial to our understanding of the Japanese workplace in
late modernity as it signifies a new process of inclusion.

Understanding of women's status in the Japanese workplace was
advanced by the analytical distinction between formal and informal structures, and indexing between inside and outside. Women's low status under
formal structures (wage, promotion, and occupational role) was counter-
balanced by their high status under informal structures (gossip, assuming

the maternal role in relation to the male worker's role as the child). This in/formal conceptual divide originated from attempts to critique male-centric accounts of the workplace. This dichotomy privileges women's perspectives, and their power is made visible. Therefore, this approach is an important starting point from which to attempt to extend our analysis of gender and status. What concerns me is that the critique borrows the same categories that men use to define women's status: women still perceive their status through the categories used by men. I appreciate that the setting up of binaries is "an instrument and a consequence of 'making equal' " (Spivak 1997: xxviii). Yet the process of freeing the subject from one perspective by switching perspectives, in other words, by creating a binary, shows in the process that "the two terms in the opposition are merely accomplices of each other" (ibid.). This general problem can be put thus:

> The opposition of male and female power is thus perceived as the opposition of semblance and actual power: man is the impostor, condemned to perform empty symbolic gestures, whereas the actual responsibility falls to women. The point not to be missed here, however, is that this spectre of women's power structurally depends on male domination: it remains its shadowy double, its retroactive effect and, as such, its inherent moment. For that reason, the idea of bringing the shadowy woman's power to light and acknowledging its central position publicly is the most subtle way of succumbing to the patriarchal trap. (Žižek 2005: 56)

Therefore, in looking for a way forward, I want to suspend the distinction between formal and informal structures, and between inside and outside. An experiential and processual approach will show that status conveys the tension between *both* formal and informal structures, and, inside and outside distinction of gendered spheres. In everyday practice in/formal structures are collapsed into each other and are complementary. Dichotomizing men from women; inside from outside; and formal from informal structures presupposes that relations between them are static, when in fact relations between them are constantly in the process of being made and remade (cf., Bakhtin 1981).

In the following chapter, by focusing once again on the experience of JCars, we will explore the redefining and legitimization of occupational role as well as the multiple ways women and men experience their positions.

CHAPTER 4

Firm Entry, Tracking System, Careers, Status Negotiation

In the previous chapter, I outlined the wider historical, socio-cultural, legal, and economic climate of the late 1990s. In this chapter, I examine how those transitions shape JCar's employment practices, workers' identities, and career trajectories. Thus, in examining employment practices, I keep to the core themes of the book, and my analysis refers to gender, status politics, and ideology. Chapter 4 thus explores the following areas of employment practice: recruitment and route of entry (via personal connections; female university and junior college graduates; male university graduates in clerical and technical positions; disabled persons); the tracking system—discussion of the clerical and career tracks (including inter- and intra-gender differences); the changing conception and experience of occupational role; conversion to the career track; women's career aspirations; women and men's talk of male careers; personality and appearance of new recruits; femininity (masculinity is discussed in chapter 6); and, the changing social perception of female office workers. Without further ado, let us explore JCar's employment practices in detail.

Recruitment

The Annual Business Plan and the Recruitment Project

The Japanese economy, JCars' business plan, and the recruitment process are intimately linked. Thereby, the company policy relating to production, marketing, and profit targets have direct bearing on the details of the recruitment project.[1] The overall policy is largely fixed and is fine-tuned according to a smaller time scale of three–five years. The outlook of the three–five-year business plan is stable and recruitment targets are roughly outlined to coincide. The annual business plan is then drawn from these mid-term plans. This annual plan is based on a close study of short-term economic projections. Close attention is paid to industry specific economic indicators. This data is gathered and analyzed between December and March by the management administration department, which together with the production administration department, set the sales plan and production plan. This is forwarded to the personnel department that analyzes the recruitment needs for the coming year and set realistic recruitment targets in consultation with the accounts department that deals with the profit plans (*rieki-keikaku*) for the company. This annual plan is then adjusted midway during the business cycle (beginning in April in Japan).

Firm Entry via Personal Connections: Quality of the Worker

JCars recruits female university graduates predominantly through personal connections (*kone* or *enko*) to existing staff or though the staff of affiliated companies.[2] The female recruits from junior college backgrounds are exempt from this selection procedure. They apply through formal routes (see below). Although men are also recruited through personal connections, as many of the older managers over 40 are, usually through relatives, the method is more prevalent among recruitment of female university graduates. This is mainly due to the prolonged recession. For this reason, I focus my discussion on women in this section. Rodney Clark (1979: 199) has shown that recruitment via connections generates and maintains cohesion among group members. This account of JCars shows how opinions on hiring via *kone* vary widely.

Through the duration of her employment (12 years), Sakamoto-*shunin* has observed the longitudinal changes to the personnel department's policy toward women's employment. Sakamoto makes the general point that "no one considers women's employment in problematic terms" (女性の雇用を問題として考えている人はいない). Since the numbers of women recruited through personal connections increased as the Japanese economy declined (since the mid-1990s), Sakamoto interprets this to mean that the company does not care about the quality of female university graduates they recruit. Sakamoto said this is true of other companies as well: obtaining a position in large companies is difficult for female graduates because they were hard-hit by the company cutbacks on recruitment. However, the problem is with the attitudes of the younger generation of female workers themselves. Ironically, the quality of younger female candidates as workers reflect the relative ease by which the recruit could enter the company. In other words, according to Sakamoto, "the calibre of new recruits had fallen" (新入社員の質が落ちた). These women give the impression that they consider work as an extension of school. They have not adapted their behavior or their expectations to a working environment requiring self-discipline (*kejime*). They do not appear to be motivated at work. Sakamoto's comment brings to mind the OLs in Jeannie Lo's (1990: 13) study of Brother Industries, Nagoya

Prefecture: these women are not concerned with long-term careers, but work in large companies to increase their marriage prospects, and also to save up for their wedding dowries. But this negative impression of younger workers is partly due to firm entry having become less competitive now than it was in Sakamoto's generation.

Conversely, Ashimoto-*shunin*, younger than Sakamoto by four years, argues in favor of recruiting via personal connections. It is a pro-active measure, effective in its ability to secure the best workers (*yūsei-shain*) in adverse economic conditions. Moreover, "it is proof of a good company when it is asked to consider a candidate via personal connections" (縁故で入れてくれと言ってこられる会社は良い会社という証し), hence it is a symbol of the company's reputation. Equally, Ashimoto is critical of this method. "The concept of *enko* itself has no regard for women's integrity" (縁故のコンセプト自体女の人を馬鹿にしている), as if the woman is incapable of finding work through merit, or by her own volition as an independently minded female rather than as a daughter or relative of a senior manager.

The workers recruited via personal connections might incur negative judgment, as some requests cannot be turned down, given the potential to weaken personal ties at the top of the organization or between related companies. Igarashi-san explains that when the personnel department consider taking on a female graduate through personal connections, they take into account the size and reputation of the company and the status/rank of the individual who makes the request (only senior managers and above are considered). From the perspective of the personnel department, *kone/enko* "guarantees" the quality of the new recruit, albeit based on a simple correlation between the candidate and her source of connection. However, many (both *kone* and non-*kone* entrants) believe this logic lacks persuasiveness. Therefore, workers such as Sakamoto look negatively upon workers recruited through *kone*.

Thus, *kone* recruitment is understood by workers as a means by which the personnel department—at the level of the department as a unit in competition with others, or for the good of the company—establishes and maintains its own connections to external firms and departmental status within the company. Negotiating the hire of a worker by *kone* is also seen to be a way a manager involved in the process competes to better one's status vis-à-vis others in the department. It also ensures that feeling-ties bind the worker ensuring that they stay on at the company.

In Japanese society, as implicit prejudice against women with postgraduate degrees exists, for "overqualified" women such as Mori-san, the only foreseeable option in gaining entry into a large, reputable company is by using personal connections to her advantage. Although women with postgraduate degrees are increasing, they remain in academia or apply for technical jobs (*gijutsu-kei*, e.g., engineering). They are still a minority among women seeking clerical work (*jimu-kei*) in large traditional companies. Mori said her female friends with postgraduate degrees from top universities in Japan rarely succeeded to the interview stage. And if they did, many are asked the same question: are you willing to accept the same salary as a male graduate with a bachelor's degree?

This exclusion of overqualified women from fair competition in labor markets can be explained partly by the downturn of the economy during the 1990s, and as the comparison with male recruits with bachelor's degrees shows, there is no fair place for women of her position. The company hierarchy and its ranking system by education, age, and tenure have not made provisions to account for the status of female individuals who have attained additional educational qualifications. Among women's qualifications, the two-year junior college diploma and four-year bachelor's degree are norms through which women are ranked, and Mori's master's degree is only as good as a bachelor's degree. Therefore, qualifications that exceed the norm become a weak signifier of official status.

Does *kone* entry reinforce the company view of women as an inefficient form of investment? Does it reflect a traditional gendered expectation, whereby *any* woman is assumed to retire from the company earlier than men, therefore making the investment of training in women wasteful? When the economy is slow and company profits are down, the company would see some benefit in recruiting women through personal connections, who might bring advantages for the company: strengthening intercompany or interpersonal bonds within the company. Indeed, business dealings are often reinforced by exchanges of favors between companies (Rohlen 1974: 68). An awareness of this system of bartering might/not undermine the self-confidence of the *kone* entrant; this depends on how the entrant interprets her role. Rather, what might be particularly harsh is the knowledge of negative impressions that other workers without *kone*, both male and female, often harbour about them—this discourse generalizes *kone* entrants as individuals of low capability, unable to enter the company without assistance, it is an unscrupulous attempt to sabotage reputations. Hence, at different levels of interpretation, Ashimoto's criticism of the recruitment of women through personal connections as damaging to the integrity of women makes sense.

On this matter of job attainment via personal connections, we locate a different view based on interviews and everyday conversations: the women who entered via *kone* do not feel excessive anxiety over their status, and the same can be said of male employees. This is a matter of shifting from a group-oriented perspective that draws heavily on how co-workers view the individual to a self-oriented one. This allows us to understand what individuals feel and value about themselves. Furthermore, this lack of anxiety concerning the link between personal connections and status can be explained by reference to research on economic life and labor markets: for social network theorists, informal job search mechanisms (use of personal contacts) represent what individuals can actively do themselves in finding employment, while formal mechanisms (advertisements, employment agencies, and job fairs) are initiated by organizations (Drentea 1988). Therefore, *kone* recruits are not merely passive pawns in the world of business; they may certainly play a part in the larger system but of their own volition. Moreover, cross-culturally not only do we find the use of personal contacts in recruitment commonplace, but they are crucial to getting a job. This is because networks of social contacts provide access to valuable labor market information that is unavailable through more formal means (Granovetter 1974; Kanter 1977). In his survey of professional, technical, and managerial employees in the United States, Mark Granovetter (1974) found that 56 percent of people found jobs through personal contacts; and, Patricia Drentea (1988) found that over 73 percent of men and women in her study gained access to their current jobs via personal connections. Such studies among others show that the advantages of using personal networks accrue to both individuals and firms in that (1) the use of social networks result in jobs of higher satisfaction, earnings, and prestige; and, (2) the information contained in social networks creates better matches between job seeker, future colleagues, and organization.

While social network theorists suggest that entry into organizations via personal connections is commonplace in many firms in the United States, at JCars where informal recruitment methods is a matter of routine, nevertheless we note a somewhat problematic link between connections and status. For the worker with *kone*, *kone* becomes an inescapable marker of status. In the unenlightened discourse that unfolds, employees lacking connections frequently gossip about the networked individual's reputation, by juxtaposing the strength and weakness of *kone*—dependent on the source of the connection—with the individual's performance at work. Ironically, such comments elucidate the distinct disadvantage felt by individuals who themselves lack connections to influential networks, particularly as early labor market success tends to have a cumulative and therefore beneficial affect on career trajectories (Granovetter 1995).

Thus, with regard to status perception, *kone* comes across as both source of empowerment *and* encumbrance, contingent on whose perception (those with or without *kone*) and kind of judgment (positive or negative) is integrated into the formulation at that particular moment. Finally, from the frequency with which the topic of *kone* is evoked in conversation, we might infer two facts: first, the degree to which *kone* as a marker of status is imbued with ambivalence and uncertainty, and second, how status is in constant need of legitimation.

Recruiting Female Junior College Graduates

Due to the economic downturn, in addition to placing widespread job advertisements and mailing of company information booklets, the personnel department sent out recruitment details to two junior colleges in the Kansai area. This is a form of quality control to guarantee higher calibre candidates. As they are recruiting in small numbers (under 10 candidates were to begin in April 1999), they concentrate on recruiting from the two top private junior women's colleges in Kansai. This is an efficient tactic.[3]

At the first group interview, lasting 30 minutes, the candidates are asked to give a self-introduction (*jiko-shōkai*). The recruiters try to get a sense of the individual and assess their logical capability and suitability for supporting roles. At the second round of interviews the candidates meet the section head of the recruitment section and the head of the personnel department. On 7 July, offers are made to the candidate by telephone (*naitei*) and on 1 October, the official offer is put to them in writing.

Allocation to Departments: Suitability

The personnel department has one week to consider the allocation of each new recruit to a department. This phase is necessary, as Japanese companies do not recruit to fill specific positions. Although this matter is important to individuals, the personnel department tends to be economical with their time, given the high turnover of women. A request containing the number of vacancies is sent from each department to the personnel department, and from this up-to-date information the allocation begins. For women, the personnel department tries to fit its impression of the women's personalities gathered during the interviews to the character of the department. They also consider the suitability of the recruit in terms of their existing skills, for example, recruits with linguistic skills have a high chance of being assigned to the overseas department and those with an aptitude for maths might be assigned to the accounts department.

At the interview, applicants from junior colleges are asked to provide preferences for the allocation to departments. They are also asked a question that is typical in many Japanese firms: "support roles exist, but how would you feel about it?" (補助的な仕事とかあるけど、どう?). Generally, the applicants from junior colleges do not have clear answers, as they seem not to have thought that far ahead. But the personnel department asks these questions as an act of formality. Again, this is a reflection of the arbitrary nature of the allocation process. In fact, the personnel department has the first pick from the pool of new recruits. Igarashi said, "rather than becoming allocated to an elite department, women are happier to be in a place that has a pleasant atmosphere" (女の子はエリートコースの部署に配属されるより居心地のいい所の方が幸せなんじゃないかと思う).

In terms of allocation, *kone* entrants incur restrictions because some departments deal with private information that is harmful to the company if leaked. For example, if the recruit's original connection to the company is through an employee of a client company, the accounts depart-ment (*keiri-bu*), secretaries to the director (*hisho-shitsu*), and the personnel department (*jinji-bu*) are unsuitable.

Recruiting Female University Graduates via Career Track

This section refers to female university graduates joining in April (*teiki saiyō*). Prior to 1999, female university graduates applied through the cleri-cal track and, after one year, based on the results of the career track exami-nation, could be promoted to a career track position. Men automatically enter through the career track. However, following the legislation of the 1997 EEOL (enforced 1999), new female recruits starting work in April 1999 are permitted to apply for positions in the career track immediately at the beginning of their careers.

For the recruitment year starting April 1999, the initial applicant pool was 300, which consisted equally of male and female applicants for career track positions. While 30 remained in the second stage of selection (the first inter-view) after passing the written examinations in the first stage, only 2 women made the third and final stage of selection (the second interview). Both Igarashi-san (female) and Tomiyama-*shunin* (male; monitor of JCar's recruit-ment policy for consistency with EEOL guidelines) said this is a deliberate tactic of exclusion: the company considered women for career track positions, yet systematically rooted out female university graduate applicants through-out the round of interviews. This way, outwardly the company appear to conform to the new legislation against gender segregation as stipulated in the 1997 EEOL even though few women are offered positions.

In truth, the company does not put women in career track positions for cost reasons. The personnel department reasons that women recruited straight into career track positions present a financial risk not worth taking. This includes the probability of women becoming pregnant at some point in their careers, but also the view that they might be unable to cope with the respon-sibilities associated with the position. However, this opinion about women's aptitude and job performance is entirely speculative, and, the assumption of retirement on childbirth is rooted in a traditional patriarchal view of women (although in truth *some* women do decide to retire on marriage and child-birth). And if the performance of existing female workers is a measure to go by, we might say that the personnel department is overly cautious.

Recruiting Men

Male recruits are considered for career track positions (there is no alterna-tive). The divisional criterion among men at JCars is not one of status (as it is between women) but of work type: between clerical and engineering/

technical work. This distinction does not apply to women's work. This is because, although women do apply for engineering/technical work, such women are still a minority among men who make up the majority of applicants for engineering/technical work. At entry, all men are on the same wage scale despite this distinction in job type. For women, during the first year of employment all are on the same wage scale, and after one year, wage differentials become induced by converts to the career track. However, career track women continue to receive lower wages than men of their cohort.

The recruitment method for male candidates is diverse and dynamic. Ties between university and company are strong. Male applicants find out about the company in seven ways. The strategic (*senryaku-teki*) recruitment method consists of the following: (1) through a private company that has lists of student names at each university, JCars can request application forms to be sent to the universities of their choice (e.g., for clerical work, 24 universities; and for technical/engineering work, 16 universities are targeted); (2) directly mailing post cards with company advertising to students' private residences; (3) sending company open day invitations to universities, held twice during April; and (4) sending workers to their alma mater to speak to graduates in person about working for the company.

The company's inactive method of recruitment where the candidate is left to seek information about the company includes placing an advertisement in student oriented employment magazines with a national circulation (the names of publications are omitted to maintain JCar's anonymity); JCar's website, they can email to request further information; adverts and press releases in the national press or business journals (again, names omitted). Taniguchi-*kachō*, a section head of the recruitment section in the personnel department, notes that some methods are more costly than others.[4] But all are necessary measures to recruit the best candidates. Although the company is a leading player in local and global industry, their products are not consumer durables, hence the JCar name is less visible to recruits and the general public than famous household names such as Toyota, Honda, Sony, or Panasonic. JCars goes to these lengths to recruit men in order to compensate for this lack of visibility.

Separate procedures apply for recruiting male applicants for technical/engineering work. Male recruits for the technical/engineering section are only recruited through the nomination method. This is a traditional method of recruitment for most large Japanese companies. The recruitment team has established relations with professors at select universities and the professors nominate their best candidates. The candidates take entrance examinations, but are interviewed once only as they have the personal backing of the professor who is trusted by the personnel department. In addition, this method is fail-safe for JCars, as once the offer is sent out it cannot be rejected by the candidate. An underlying trust between the company and candidate supports this recruitment method. This rule is similar to the nomination method for the recruitment of female junior college graduates as mentioned above. University professors do not nominate candidates for clerical type work. After the entrance examinations, these men go through three rounds of interviews. The personnel department looks for men who exhibit qualities of enthusiasm and assertiveness (*sekkyoku-teki*) supported by clear career aims.

Exceptional workers are recruited steadily and both male and female recruits enter the company at irregular times throughout the year (fewer females reflect the overall proportion of men and female employees) and vary across a wide age range (from recent graduates to middle management). Some are head hunted from those already employed at other companies. JCars places A4 size advertisements in national newspapers and a full page in recruitment magazines (again, name omitted) consulted by men seeking career changes. JCars also asks employees to use their personal connections with people in other companies to inquire after employees looking for a job hop. As workers visit their university professors when they become dissatisfied in their current jobs, the personnel department keeps in regular contact with the top universities that interest them. Usually technology developers have a tendency for this, as their skills are transposable across various companies and sectors. In addition, JCars contacts job centers run by each prefecture (*shokugyō-hantei-jo*). They incur no additional cost for this service as it is a state-run agency (*kikan*).

The personnel department goes to tremendous lengths to recruit the best male candidates. It is tempting to draw a simple correlation between the recruitment budget allocated to male candidates and perceived value as employees. Despite (and perhaps due to) the large budget made available for the selection of male candidates, women are given lower wages than men and consigned to a structurally lower position under the economic logic of cost minimization. At the time of this study (1998–9), although the 1997 EEOL guidelines were being followed at JCars, more can be done to correct the imbalance that persists in the recruitment and hire of women.

Disabled Workers

The company's employment of disabled workers is regulated by law: *chiteki-shōgaisha-koyō-sokushin-jirei*, which aims to normalize the presence of disabled workers in workplaces:

> Normalisation . . . aims to create a society in which people with and without disability help each other to live in the community with vitality and richness in heart, [the Department of Health and Welfare for Persons with Disabilities]

will promote the independence and social participation of the disabled. (Ministry of Health, Labour and Welfare 2004)

The number of disabled individuals in Japan, over age-18, classed with the heaviest degree of disability, number 160,716 in 1995 (Osaka Prefecture Employment Department 1998: 3). The law encourages private sector companies (employing over 63 regular employees) to employ disabled workers to meet at least 1.6 percent of the company's total workforce; this figure was raised to 1.8 percent in July 1998. Higher demands are placed on public sector organizations at national and regional level where 2.1 percent of their staff must be recruited from disabled persons. At Prefectural level, in the education sector in particular, the recruitment target is 2.0 percent. These rates remain stable in 2008 (JEED 2008).[5] Official statistics compiled in 1997 by the Japan Association for Employment of the Disabled has shown that out of the 5,134 companies surveyed 48.9 percent under-recruited disabled workers. The law penalizes companies by enforcing the payment of compensation fees per employee they under-recruit annually (¥60,000 in 1998; ¥50,000 in 2008). This can hit hard on companies trying to manage cost cuts. However, companies employing under 300 employees are exempt from the payment of penalties. On the other hand, if companies employing over 300 workers manage to exceed the 1.8 percent baseline, the company is reimbursed by ¥27,000 per person they over-recruit. Likewise, if companies recruit workers diagnosed with clinical depression they are rewarded ¥30,000 per month per worker. In calculating levies, employees with severe disability count as two people rather than one.

Let us place Japan's employment measures for disabled people in cross-cultural perspective (see the Japan Institute of Labour Policy and Training 2008: 291–2). Both Germany and France like Japan (although the respective rates and minimum firm size vary), set targets to firms employing over 20 people to hire 6 percent of employees from among the disabled. If employers fall short of this target, they are required to pay compensation; the levy paid by companies is administered by the local government in Germany and by Agefiph, a private sector organization in France. In the UK and United States, it is illegal to discriminate against disabled workers in access to employment. The UK's Disability Discrimination Act 2005 promotes civil rights for disabled people and protects disabled people from discrimination; the act enables the government to set minimum standards with regard to areas of everyday life effecting disabled people such as employment; education; access to goods, facilities, and transport services; health care; and the buying and renting of property (DDA 2005). In the United States, the Americans with Disabilities Act of 1990 (ADA) prohibit discrimination in the employment of disabled individuals. This law is enforced by the U.S. Equal Employment Opportunity Commission,

State, and local civil rights enforcement agencies that work with the commission. In 2008, the definition of "disability" was changed in the ADA Amendments Act (effective Jan 2009). While the act retains the ADA's basic definition of "disability"—an impairment that substantially limits one or more major life activities, a record of such an impairment, or being regarded as having such an impairment; the act changes the way that these statutory terms are interpreted (see Amendments Act of 2008).

In Japan, the provision of support toward the employment of disabled people is centralized by the Ministry of Labour, and the Japan Association for Employment of the Disabled (JAED), like the Employment Promotion Corporation, is one of its specialist organizations, it has offices in each Prefecture. There are many agencies and facilities for vocational rehabilitation created and run at national and local government level, not to mention those run by the private sector. JAED was set up to promote the understanding of government legislation among employers and disabled job seekers; therefore one of its functions is to introduce disabled persons to potential employers.

Turning now to JCars, the two disabled workers I met were white-collar career track workers in the IT department, but based in separate offices. The majority of workers with disabilities work in the manufacturing plants. At JCars white-collar workers with disabilities are rare, although the company tries to recruit a small number every year. Like other organizations, JCars failed to meet the legal recruitment target of disabled workers. This is because the procedure is complex and labor intensive compared to recruiting non-disabled employees. Moreover, larger firms must recruit greater numbers to meet the 1.8 percent target, which is fixed, not relative to firm size.

Taniguchi-*kachō*, the recruitment manager of the personnel department, attends the annual open day held over a one-week period at JAED. A JAED counsellor performs the role of the go-between to introduce potential workers to Taniguchi. During this week he looks through numerous CVs provided by the counsellor. The potential worker's educational history and achievements are not critical to his or her employment prospects (this is in stark contrast to the importance placed on the educational qualifications of non-disabled candidates) but suitable candidates are not easily found.

Ishii-san is a new recruit who I met at the new recruits seminar in April 1998. His peers talked in their little cliques during breaks and avoided communicating with him; they looked past him. If he felt excluded from the group (as I did), he did not show it. In between sessions, I moved to a seat next to him. I didn't know Ishii was deaf until I tried speaking to him. His disability is imperceptible. We wrote messages to each other on a pad passing it back and forth during some breaks. He is based at a subsidiary in

Tokushima (far from Osaka) and is the only disabled white-collar worker recruited in 1997 to begin work in 1998.

In one session we watched a video about minority political rights, identity, and history as part of *dōtoku-kyōiku* (moral education). The managers from the personnel department also listed real experiences of discrimination against *burakumin* (untouchables) and Korean descendants that took place in factories and subsidiaries at JCars. In spite of this, our peers viewed Ishii and me suspiciously as we communicated.

Karen Nakamura's (2006: 142) fascinating ethnographic and archival research on deaf identity shows that during the 1990s, only a minority of deaf school-leavers went on to university-level education. Many enter the clerical track in white-collar workplaces or take blue-collar jobs. Nakamura says deaf identity collapses disabled and ethnic minority groups' experience of status politics (like Japanese-Brazilians, and second and third generation North Koreans), all of whom struggle to gain a space for identity construction and political power (pp. 159–62).

How would Ishii experience work at JCars? I was unable to remain in touch with Ishii in part because his office was physically distant from mine. Regrettably, therefore, I was unable to fully appreciate or understand the practical consequences he faced in working at JCars as a disabled worker. We will meet Kubo-san, the other disabled worker in chapter 6.

As a final point in this section let us consider the implications of the following statement: JCars regards its efforts as a "duty to society" (*shakai-teki-sekimu*) and wish to increase the numbers of disabled employees. Tomiyama-*shunin* in the personnel department raised the following point to explain why JCars finds it difficult to meet the recruitment targets and fulfill their duty to society. It is that JCars cannot recruit disabled workers in large numbers because heavy metal products are handled in the factories; disabled workers are not employed to work at a single site in large numbers to avoid compromising their safety. Tomiyama says large companies in the manufacturing sector producing smaller consumer goods can provide a safer and better environment for the employment of disabled workers. Yet it is well known that large companies dealing in the assembly of small parts set up factories specifically to meet legal disability recruitment targets, for example, the Panasonic factory in Katano City, and Kansai Electricity's facility near Nankō harbour (the government provided the building). I did not visit or study the facilities above but I am not convinced that duty to society can be met by recruitment in numbers; surely it is the quality of the work environment evaluated from the disabled workers' perspective that deserves careful attention.

Although JCars does not endorse this form of "sheltered employment" that segregates disabled workers into a separate sphere—away from the regular competitive labor market, it merits discussion because the practice appears inconsistent with the government endorsed meaning of normalization that aims to integrate disabled workers into workplaces and society (see definition of normalization at start of the section). Some significant questions come to mind. Does the herding of disabled workers into one site make supervision easier? Is it a safer work environment for them because facilities and machines are fully adapted for a disabled person's use? Or is this segregation simply a labor market logic that operates above all of us—that is, cost-savings gained through low remuneration. Will disabled workers not lose their dignity by working in such segregated conditions? Does it not prevent disabled workers from acquiring valuable interpersonal and life skills that enable them to function independently in society? How does the public view this segregated employment of disabled workers and how do their views affect the disabled person's self-esteem and place in society?

Disability related employment is a challenge to most employers in the EU, United States, and Asia (Rojewski 1999). Specifically, with regard to sheltered employment of disabled people, in the United States, this practice is criticized for warehousing workers rather than providing skills development (p. 248); in China, where there are 50,000 government-subsidized factories (tax exempt), there exist only two avenues of employment for disabled persons capable of working—either sheltered employment or self-employment (p. 250); in Japan, employment in sheltered workshops (refers to rehabilitation in this context) increased as conversely access to open labor markets decreased, and although sheltered workshops assist in obtaining competitive employment outcomes, the annual placement rate of disabled individuals in competitive markets remains at a low 2 percent (ibid.); in Germany, for people with cognitive or learning disabilities sheltered workshops are the only employment option, and these offer low remuneration (Waldschmidt and Lingnau 2008). Furthermore, Stephens et al. (2005) have shown that working in an open (competitive) labor market as opposed to a sheltered or supported workplace enhances the adaptive skills of people with developmental disabilities, and the newly gained skills add to their success in communal living.

Thus, rather than rushing to meet recruitment targets one hopes that all companies will reconsider the meaning of normalization with due care. This requires a shift in perspective in how the employment of disabled persons is viewed. The current thinking is company centric: the disabled person who is most like a non-disabled person must be recruited so as not to disrupt the existing environment, that is, those whose disabilities are imperceptible like Ishii and Kubo. Although maintaining the environment is clearly important, and this is not unique to disability, as, in this regard, rigorous selection criteria also apply to regular workers; nonetheless, moving from an attitude of reluctance to acceptance is of equal import. What is needed from companies is an openness and greater belief (always entails risk) that allows the disabled person to be brought to the fore: by accounting for the disabled person's individuality, in seeing what the disabled person adds to the workplace. Moreover, firms can be encouraged to embrace greater creative and pro-active thinking with regard to their role as skills providers. If the company wishes to maintain its community ethos, is it not vital that they foster an environment enabling disabled workers to live and work inclusively?

The Tracking System

Experience of Positions in Clerical and Career Tracks

一般職は上司に手伝ってもらえる、途中であきらめられる。

[The clerical track worker can receive help from the boss, and give up halfway.]

総合職は責任感をもって最後まで自分でやる。

[The career track worker assumes a sense of responsibility and pursues the task to the end on one's own.]

("What is the difference between the clerical and the career track?"
Mori's reply in her interview for conversion to the career track, March 1998)

At JCars employment status ranges from regular to non-regular (i.e., temporary).[6] These forms of employment produce a hierarchy through which status is determined, and generally this correlates to gender, age, year of entry, educational background, and presumed ability. The recruitment policies discussed in this chapter concern regular employees (*sei-shain*) who have non-contractual agreements with the company: the length of employment is undetermined and it is assumed at the outset that workers will stay with the company until retirement. I focus on these workers as the majority of the women and men at the head office are recruits of this category.

There are a few temporary workers (*ichiji-in*) recruited on a contractual basis through recruitment agencies (*haken-shain*). These female workers in their late twenties and thirties are, as Heidi Gottfried (2003) notes, expected to blend into the informal organizational culture without belonging to the formal organization. At JCars *haken-shain* are quiet and keep to themselves, without mixing with the rest of the female workers during lunch hours or after work. Some regular female workers look down on them, treating them as an inferior class of worker, but male managers are impressed with the quality of their work and the enthusiasm they bring to it. In terms of appearance, they wear more make-up and had lighter colored hair compared to regular workers.

Igarashi-san has worked at JCars since graduating from one of the top national universities in Japan. She joined the education section of the personnel department in 1993, and after three years requested a transfer to the recruitment section.[7] In her fourth year, the department granted her the transfer. Igarashi was recruited on the clerical track. Many women questioned her decision to remain in the clerical track as she had the experience and elite educational background to make her an exemplary career track candidate.

Igarashi said "it would be awkward to become the first [to convert to the career track] as no other women in the personnel department are on the career track" (人事に総合職の人はいないから自分が一番になりにくい). She is fearful of disrupting social relations with women in the department: "afraid that the other women might change their attitude" (他の女の子達が態度を変えるかもしれないから恐い) toward her if she rose in official status by joining the career track. This fear of other women and destructive competition between female colleagues is also experienced among Western business women, where successful women are labeled "arrogant," "dominating," "unfeminine," or as "the bitch" (Hite 2000; Tannenbaum 2002).

Igarashi is aware of other departmental managers who encourage women to convert to the career track, but this is not the case in the personnel department. Speculatively, she attributed her manager's lack of enthusiasm to a patriarchal attitude: "men probably think that a woman, after all, should just be doing support [work]" (男の人は, 女の人は所詮サポートしとけばいいと思っているだろう). Yet, she was "now doing the same work as a male worker" (今は男の人と同じ仕事をしていると思う) and "thinks she is being evaluated accordingly" (評価はそれなりにされていると思う). As the managers perceived her as equally capable as her male colleagues, she said this encouraged her to be more conscientious toward her work. Igarashi said, "not being on the career track doesn't mean that I take indulgence in my position and place limits on what I will do, or that I think of leaving my work to other people" (総合職になってないから 甘えがあってこれまででいいとか 他の人に任せようとか思わない). Igarashi was also reluctant to change to the career track "...because, it is already difficult now and I will be troubled if I am given more work" (今でも大変なのにこれ以上任されても困るから総合職になろうと思わない). From Igarashi's comments the difference between women and men's work in this department seems to be based on the amount rather than on the difficulty, content, or kind of assignment.

Tamaki-san graduated from a private women's four-year university and is a clerical track worker in the personnel department. She has been working for two years. Tamaki's evaluation of women's work and the degree of responsibility given to women are similar to the comments made by Igarashi above. Tamaki said the main difference between the personnel department and other departments is that in the personnel department, all women are on the clerical track, yet they are given responsibilities and work equivalent to a career track position. With no boundaries between the clerical and the career track, status distinctions between women are difficult to define (部署によるみたいだが 人事では一般職の人でも総合職と同じことをしている。だから一般と総合職の区分は難しい). Thus, she implies that all women are of equal status in the personnel department.

In the overseas department, the type of work given to women is clearly defined and delineated, reflecting the division of women into clerical and career track positions. Tamaki said some women in the clerical track in the overseas department are given supporting roles, such as sending faxes and making photocopies for others, whereas in the personnel department, men and women alike attended to their own faxing and photocopying. The personnel department has the first choice in the selection of recruits. In so doing, the personnel department is able to acquire the best candidates and thus Tamaki reasoned that women entering through the clerical track could be given difficult jobs equivalent to the career track jobs.

Tamaki's senior female "sister" in the personnel department who retired during my fieldwork was encouraged by her managers to apply to the career track. She had refused and remained in the clerical track because the conversion would not make any difference to the type of work she was given: she was already entrusted with a demanding workload. Tamaki mentioned Ashimoto-*shunin* who converted from the clerical to the career track while assigned to the personnel department. This occurred before Tamaki joined the company. Thus, Igarashi would not have been the first woman in the department to convert. And it was the managers in the personnel department who, in fact, recommend the upgrade. Ashimoto was relocated to the Tokyo office for a period of two–three years.

As it turns out, for Tamaki, the possibility and reality of relocation was the main factor distinguishing the clerical from the career track rather than work content. This is true of the overseas department as well (e.g., Sakamoto-*shunin* relocated to a subsidiary in the EU). However, no other women on the career track from the overseas department have been relocated since Sakamoto's assignment. The managers in the overseas department said this is because the directors of JCars did not want to send female candidates overseas, as it is more costly as they have to provide housing benefit on top of an expat's wage. Also worry over the women's welfare on account of their gender adds to their concern. It is less a matter of which countries are considered the most threatening in terms of women's welfare than about women being seen as vulnerable creatures in need of patriarchal protection both inside and outside of Japan.

The management limits the promotion of female workers as a cost-saving measure. The paternal concern for a woman's physical and mental safety (genuine or feigned to cover discriminatory views, depends on the point of view) boils down to this as well, for in the end, in the event of some unlikely disaster involving an expat female worker, the company would face real legal repercussions. Further, by protecting female workers who go abroad on account of their feminine vulnerability, men are able to maintain their positions by controlling the social structure of the organization, as well as protect the symbol (women) of the nation-state (Meillassoux 1981). Table 1 summarizes the differences between clerical and career track. On a comparative note, these differences parallel fast-track and non–fast-track

Table 1 *Summary of Differences between Clerical and Career Tracks*

The Clerical Track	*The Career Track*
① Initial Allocation and Subsequent Conversion Criteria	
Junior college graduates and university graduates	After one year, only university graduates are allowed to convert
② Job Content: Official Definition	
• Less responsibility • Lighter workload • Lower wage • Fewer job rotations • Promotion limited to lower level positions	• Increased wages, levels of responsibility, workload, and overtime • Involves possibility of relocation • Wages for women remain lower than male colleagues of the same cohort
③ Real Experience: Point of View	
• For the majority of women in the clerical track, the workload is very similar to the career track • Some think work in the clerical track is easier than in the career track • Some see themselves as doing the same work as men	• Some women in the career track perceive their workload and level of responsibility on the job to be greater than that of clerical track workers • Some see themselves doing the same work as men • Some think clerical track workers are hard working, doing the same jobs as them
④ Implications for Status Differences	
• On account of experiencing a similar workload to career track, some women saw themselves as having equal status to career track • These women did not think educational background should be a marker of status difference, or that status should be determined by work performance	• Status on career track supersedes status based on tenure • Some career track workers see themselves as having more status than clerical workers: this means that their status is distinctive insofar as management recognizes their higher educational qualification

Note: The complexity of experience, depending on the point of view taken, produced a blurring effect on the boundary between clerical and career track positions given in ②.

personnel in the Ministry of Foreign Affairs (*Gaimushō*) and in the British Foreign and Commonwealth Office.

Restructuring: Job Equality between Clerical and Career Track—Occupational Role

Let us account for the structural reasons of how this experience has come about. Describing the effects of restructuring on the personnel department, Kishi-*shunin*, a male worker in his early thirties, says that numbers decreased from 30 to 29 workers in the past 10 years. As a direct outcome of this, women have been given work with more responsibility (十年前より人事部の人数は減っている。私が入ったころ三十人いたが今は二十人、だから女性に対して責任を任せる仕事が増えた). Ota-san, a female worker in her late twenties, mentions how "there was a regular turnover, but with the current personnel cut backs it is no longer this way. [And] the amount of work increases annually" (前迄は一人やめると一人入ってきていたが人員削減で今はそうではない。仕事の量は毎年増えていっている). Kishi also notes that computerization decreases headcount in tune with falling amounts of paper work, and this fall in demand for menial labor affects women. Conversely, through economic necessity but also on merit, the personnel department utilizes women's skills to make up for lost numbers. Meanwhile, increased responsibility and workload are experienced by women. And these changes are linked to women's shifting status in the workplace.

To examine this claim in perspective let us look at the overseas department, which is three times the size, and of a different character to the personnel department. The overseas department has the greatest number of female career track workers in JCars (8 out of the 10 university graduates in the department). The managers are supportive of women's careers, give women challenging jobs, and both men and women say the atmosphere is conducive to women's long-term careers. As a large department, they have a greater need for support roles to fulfill various administrative functions and temporary workers (four women) perform these mainly menial tasks. In addition, there are a handful of women in clerical positions who perform similar menial work (but they are a minority). And, again, there are clerical track women, who for personal reasons did not want overtime work or a long career like Igarashi-san in the personnel department, or could not convert to the career track while performing career track jobs because they are junior college graduates. Therefore, in the overseas department compared to the personnel department, status distinctions between women are more pronounced and intricate, since women's experiences of status and work are equally varied.

At the level of experience the difference between clerical and career track jobs is difficult to formulate in any concrete way because the company practice itself is haphazard and further differs across departments. The track system is then to the company's advantage, as the company minimizes labor costs by keeping women in clerical track positions while giving women the same work content as career track workers. This works perfectly, as many women on the clerical track are not prepared to deal with an increased responsibility equivalent to the career track or a man's, but they require recognition and respect for their ability and competence. This creates a setting wherein companies can entrust women with jobs of increased workload, without adjusting for women's wage or status in the organizational hierarchy.

What are the implications of this blurring of the clerical and career tracks? We noted that consensus holds together the views of women of the personnel department: that, work as clerical track workers is equivalent to career track work. This affects our understanding of women's status on four levels. First, they construct equal status that functions to mask differences in educational background among clerical track women. Second, they elevate their status to equal that of career track women, thereby masking their difference from, and removing competition with career track women. Third, they claim elite status as a department, although other departments take on career track women. And, fourth, they become equivalent to men as they claim equal footing based on the type of work performed. The large majority of men in the personnel department affirm their views, excluding one or two. Women's status is constructed in gender-neutral terms. On this note, these women can be said to have achieved masculine equivalence. In addition, these women seem to have forged solidarity.

What is significant here is that the terms of representation and explanation of status are self-descriptions (not chosen by the researcher). Women's experience and their interpretation of status positions pose a different picture to images of women depicted objectively. The same can be said for the tracking system. Therefore, emphasis on experience shows two different but related elements. First, the tracking system does not function objectively. It operates on a subjective level because departmental difference in practice easily distorts the distinction between clerical and career tracks. Second, women's views account for their reflexive attempts at making their situation in the workplace seem more democratic in spite of the restrictions placed on them. However, we cannot overlook the fact that, although the allocation of work tasks that equal men's in complexity is an objective fact, remuneration takes the form of evaluation and appraisal of discursively constructed status, not via objective measures in terms of increased wage or promotion. Where does this leave women?

While I argue that social structure viewed from a purely official perspective cannot fully account for women's status, we must also acknowledge the flaw (or selective application) of this experiential and interpretive approach. Although women in the personnel department do not allude to it, in the overseas department, where most of the career track workers are found, women are acutely conscious of wage differentials between male and female career track workers of the same cohort. For them, it is an important point, that, even being on career track does not put them on equal footing as men, in monetary or in terms of training. Unlike the women of the personnel department, they emphasize that objective criteria are crucial for assessing status vis-à-vis men. This means that the women of the overseas department are more aware of the awkwardness of their positions as being in direct competition with men (because all men are career track) but on unfair terms (lower wage and less training).

We know that gender is embedded in the structure of organizations (Acker 1990). Interestingly, gender disappears in discourses of status among women of the personnel department. Yet, gender is very visible to women of the overseas department. This difference can be explained at three levels—field data and theory: First, by the fact that the personnel department implements the tracking system, the system originally introduced to prevent gender segregation. Therefore, its members endorse this hegemonic discourse. Second, a "gender-free" society is the ideal vision of gender equality captured in the Basic Law for a Gender-Equal Society, passed in 1999 (Osawa Mari 2000: 6). Gender-free means "a society where the fact of being a man or woman has no effect on the options available to people . . ." The women in the personnel department could have assimilated this feminist discourse. Third, the tension in theorizing identity or status as discursive constructions (see Nash 1994: 70–1, 74) become crystallized, when, as in the case above, the structural factors contributing to a real experience of oppression simply vanishes. Furthermore, our case underscores Lisa Adkins' (2004: 199–203) critique of feminist workplace studies that follow Pierre Bourdieu's approach of collapsing the distinctions between objective and subjective criteria; her point is that it does not free women from gender embedded in organizational norms. Adkins suggests that perhaps newly gained forms of status only show a new process of categorization and classification taking shape in late modernity.

Discrepancy between the experiences of women belonging to two different departments also brings up the question—is gender a dichotomizing principle or a question of positionality? Both contribute to our understanding as relational models of gender have shown. With regard to the question of women as situation (Moi 2005), this example reminds us of the observation that gender is only one source of social difference among many (Rosenberger 1994, cited in Gottfried 2003: 261). We might also add that the meaning of gender as a system of differentiation contains a wide range of meanings (Konno 2000). Furthermore, assessing work and construing status in this way shows clearly how our anthropological subjects are producers of knowledge (Moore 1996). Our theories are not divorced from the knowledge that our subjects produce.

We raised many points but what can be said, then, is that, even though no woman escapes the clutches of objective forces that structure gender segregation in workplaces, experience is nonetheless directly relevant to the production of knowledge. Although experience is not problem-free either, since experience is, as Elizabeth Grosz (1994: 94) writes, "... implicated in and produced by various knowledges and social practices" and therefore "cannot judge knowledges." But in studies of gender, experience, as Lois McNay (2004) shows, is a useful relational concept—a lived relation—that helps us to overcome the opposition between seeing gender either as a structural location or as a location within symbolic and discursive structures. In all cases, then, we must bear in mind the highly situational nature of gender identity and status attainment and negotiation. Indeed, if experience shows that standards and expectations of gender equality vary among women, yet if this shift in occupational role identifies a new process of inclusion, even if the meaning of this experience is debatable and contested, we might still view this inclusion as a step forward for workplace gender relations.

Converting to the Career Track

The conversion from the clerical to the career track is highly selective. It exemplifies another facet of practices that sustain inequality principally within two sets of relations: inequality between men and women, and inequality among women. An unspoken company rule excludes junior college graduates in the clerical track from successfully converting to career track positions. The manager in the personnel department responsible for the examination and interviews in the conversion proceedings confirmed this. By keeping the number of women in the career track to a minimum, the company cuts its employee wage expenditure: monthly wages and bonuses vary significantly between female clerical and career track workers (by ¥50,000 and ¥200,000 respectively). Capable women remain in the clerical track despite showing the ability to perform well in career track jobs.

Women in the overseas department contest the injustice of this system among themselves. Merit is not assessed independently of factors such as educational status. Feigning fairness, the personnel department openly calls for all women to submit applications for conversion. However, so far, all junior college graduates failed to ascend to career track positions, despite strong backing and encouragement given by their managers to convert. The outcome discourages many women. For example, Shimada-san is 30 years old and is in her tenth year at JCars. She is sent on overseas business trips and is entrusted with above normal levels of responsibility for a woman on the clerical track. Yet, she has not passed the examination despite her numerous attempts.

For university graduates as well, the interviews are notoriously difficult. The written examination held in March tests for mathematical, *kanji* and Japanese language skills, including analogical comprehension. The interview is held two weeks later. For candidates the main challenge is preparing solid reasons for wanting to convert to a career track position. The four candidates interviewed for conversion to the career track in 1998 encountered chauvinistic attitudes from the manager of the personnel department in charge. Only one manager takes part in the interview and this manager has been male.

Mori-san was asked if she was prepared to move to a small town in Hokkaido (Northern Japan), even if she had requested a transfer to an overseas subsidiary. When she replied that (glamor and status of) location was unimportant and she would willingly relocate, with an incredulous snigger the interviewer accused her of exaggerating. Further, when she expressed enthusiasm in gaining greater specialized work knowledge, the interviewer replied that remaining in the clerical track would be better suited to building a solid knowledge of her job, than a career track position where relocation would disrupt the learning process. However, had she anticipated his retort and said that she was unwilling to relocate, she would have failed the interview. Candidates are thus locked in a somewhat infuriating, paradoxical dialectic.

Overall the very fact that the company restricts entry through career track positions to a few women, and the implicit and explicit obstacles women face in the process of conversion to the career track exemplify one of the basic causes of structural inequality. The attitudes of male managers in the personnel department, the decision makers who have the final say in women's careers, act as principal barriers to the resolution of structural inequalities in the workplace. The current economic situation that elevates cost cutting to a key issue intersects with the encultured view of women's positions in workplaces. When these views gain currency among women themselves, inevitably this leads to pragmatic responses. This reflects Konno Minako's (2000: 221–2) stance that there are numerous circumstances and mechanisms that support the rationale for gender segregation; when gender-based treatment is anticipated and proven in practice, it continually feeds into continuing patterns and practices of gendering. Similarly, Cecilia Ridgeway and Shelley Correll (2004) also argue that when hegemonic gender beliefs are prominent in a social relational context it modestly yet systematically differentiates behavior and evaluation. Furthermore, interrelations among women are another significant barrier to changes in women's status in the company. Fear and anxiety of other women run deeply. Regarding women's status positions in the workplace, inequality sustained among women has as much impact as inequality between men and women.

Sakamoto-*shunin*, a junior section head, has been working at JCars for 13 years. She was among the first to convert to the career track. She entered the company at a time when both the 1986 EEOL and the tracking system were introduced. Sakamoto described the reactions to the institution of clerical and career tracks: "severe conflicts" (対立が激しかった) ensued and she thought "women are unable to unite" (女性は団結できない).[8] Female colleagues who had longer careers than Sakamoto thought she was arrogant as she superseded them in status: the career track supersedes length of tenure as a mark of status (cf., Takenobu 1994: 43). There was some bullying (*ijime*) of career track women by the clerical track women (details not given) producing a deferential climate where the former placated the latter (*ki o tsukau*).

Theoretically, the career track is thought to put women in an equivalent social position to men and, in this regard, Sakamoto thought career track women should be exempt from menial tasks (cleaning the tea room and handing out circulars). However, she continued with these duties, fearing the reaction of the clerical track women. Sakamoto said the clerical track women downplay differences in the nature of tasks performed by clerical and career track women. However, the placating on behalf of career track workers feed this attitude. Moreover, ironically, women's status, as it appears to a minority of male workers, as workers who perform menial domestic tasks, is to an extent maintained by this conflict among women.

In her fourth year at JCars, in the late 1980s, Sakamoto was transferred to the subsidiary in the EU. Before her relocation, she performed feminine-domestic duties such as wiping desks as a good-will gesture. Paying her due was essential to ensure the continuing countersupport of her female colleagues following her relocation. Women attempt to keep other women down. We can thus note the shift in the site of status conflict from intersex to intrasex spheres. The ideological clashes in workplaces analyzed by Susan Pharr (1990) in her "rebellion of tea pourers" was motivated by unfair status differences between men and women, and set in the 1950s and 1960s. Advancing on this picture, at JCars, with the introduction of the tracking system that sharpened status distinctions between women, women targeted their complaints against other women.

True to form, 13–14 years after Sakamoto's experience noted above, the same tensions caused by differences in age, tenure, and tracking complicate interrelations among women. Company policy is largely responsible for this. The fact that junior college graduates are discouraged from career track positions where by contrast, university graduates seem to become career track more readily, incite tension whereby career track women are constantly watched and judged by clerical track women. Thus, some career track women feel embarrassed when a fellow career track woman takes advantage of her position and does not work as much as she should. Indeed, disputes are rife among career track women themselves. They jostle for status, competing on basis of legitimacy (who deserves [not] to be on the career track), by who works harder, whose overtime is longer, and who works on the more difficult assignments.

What is being done to resolve this structural inequality among workers? To address the problematic role of the tracking system, the personnel department considered bringing women on par with men by either losing the clerical track all together or by creating clerical track positions for men. Under the existing tracking system, all men are recruited into the career track (even Kubo-san, the disabled worker with a high school education and technical school diploma [see chapter 6]) on the assumption that male recruits all have university degrees. In fact, this reinforces the fundamental origin of inequality—that individuals are assessed foremost in terms of gender. Moreover, the creation of a clerical track for men would not alter the social structure in the head office. Therefore, this reorganization will redress women's status neither in relation to men nor between women.

This policy would be applied in manufacturing plants to accommodate the transition of line workers advancing into office jobs within the plant. Most plant workers are high school graduates. The main problem envisaged with the creation of a clerical track for men in the plants was the complex reorganization of the existing social structure. The personnel department said "creation and allocation of new status positions is a problem" (処遇をどう分ければいいかが問題). But one questions how female clerical track workers with junior college and university degrees will feel to be on the same track as high school leavers (even if they do not interact in the same facility).

In the end, it is easiest to maintain the existing social order: through sets of gendered oppositions. In a gender-biased organization, that is, a principally male-dominated organization, women would continue to be treated unequally, mainly due to their gender, through attitudes of complacency, reluctance, and inertia: changing the elementary structures of the organization is considered demanding and unnecessary from a management perspective.

EEOL's Effects on Women's Long-Term Career Aspirations

Ashimoto-*shunin* has been at JCars for 11 years. She worked in the personnel department for five and a half years and in her fourth year (1991) converted from the clerical to the career track. She is now in the overseas department. From the first year of her employment, Ashimoto was assigned difficult work equivalent to men's work (入社して一年目なのに男の人と同じきつい仕事をやらされた). This is before the tracking system was introduced.

While in the clerical track she felt constant pressure to prove herself to her colleagues. Unlike other women in her cohort, she had no time to take traditional tea or calligraphy lessons after work. When Ashimoto began working she expected to retire after two or three years after finding a husband through an arranged marriage; such were the social expectations toward women. (Hori-san said most women at the company marry after five years of starting work, and junior college graduates tend to marry before university graduates.)

Despite the difficulties presented by the management and tension among women, Ashimoto joined the career track. Her reasons were twofold: first, she was dissatisfied with the wage differentials between women and men at the time and, second, she enjoyed working, she was capable and wanted to continue. The head of Ashimoto's department supported her long-term career aspirations. His support was central to her decision to continue working. The EEOL also backed her long-standing presence at the company amidst opinions of men who objected to women having full careers.

From Ashimoto's experience of work in both clerical and career track positions during the time when the EEOL introduced changes to company policy, insofar as the law legitimized women becoming long-term employees, we can see that there was some efficacy to the law as it exerted some impact on shifting the social perception of women in the workplace at a general level. Her example shows that women can make the EEOL work in their favor. The general view is that the number of women showing greater commitment to their jobs has increased since the tracking system was introduced (*Nikkei Business* 27/01/97). As evidence of this, in 1995, single women between 35–40 years old made up 40 percent of all loans/mortgage applications, which is indicative of a greater number of women pursuing long-term careers (ibid.).

For career-minded women like Ashimoto, the key factor is having an understanding manager support her career aims. She criticized the personnel department's indifferent attitude toward women's career prospects. Many departments still attempt to treat female workers on equal terms to male workers insofar as their talents go, but not as long-term workers. The overseas department is exceptional in this regard. The company can cut costs in this way and still appear to treat women fairly, because the gendered division of labor in the home supports their outlook. Wives and prospective mothers are culturally valued, but this positive view of women in the home functions negatively against career-minded women in the workplace, because these values influence most women and men about the amount of work and responsibility women should take on. Therefore, aside from the problem of the tracking system, what otherwise halts the rise of women's status is their own antagonism toward long-term careers. Partly this depends on their husband's attitudes to childcare, and also the attitudes of some managers. During my fieldwork (12 months) one tenth of female workers retired, mostly citing marriage as their reason. As long as managers know that the general expectations of women are to retire early, women will be treated as the dispensable part of the workforce. However, women are prolonging their careers. When I made a visit to JCars in October 2003, four years after leaving the field, I recognized the majority of female workers in the overseas department.

Women and Men's Talk of Male Careers

After five years, differences between men begin to appear as some are promoted while others are left behind. Ability, merit, and an influential boss are the main ingredients for promotion. Intriguingly, women, although not men, speak about a man's designation on the "promotion course" (*shusse-kōsu*). Men can stay on or fall off course: initial access seems to depend on an elite educational background, then staying on depends on his job performance and handling of interpersonal relationships. On top of these assets, at times, an imagined competence for work (仕事ができそう) is collapsed onto a man's good looks. In addition, destiny plays a hand—this is magical thinking and a culturally relevant thought.

This is a woman's romanticized notion of men's careers, a view of the elite minority, and a glamorization of masculine ideals. Masculine ideals rarely correspond to real men, rather they express "widespread ideals, fantasies and desires" (Connell and Messerschmidt 2005: 838). Yet, at the same time, the observation is grounded in real life, where women closely chart the rise and fall of men based on men's speed and accuracy of work, and their relationship with their bosses and subordinates. Women watch men closely indeed.

In women's discourse on individual men's careers, educational achievement, merit, appearance, and class intersect. Tabata-san, a young man in his twenties, joined JCars mid-year. Many women were impressed by him, and when he first joined he was the center of attention. He is tall (about 180cm), slim and good looking, graduate of a top private university in Tokyo, with entrepreneurial experience in setting up his own company. He is a success. Yet, after a few months, based on fresh information, his status was re-evaluated, which then declined in the eyes of some women. They were disillusioned that he was not an *o-bochama* (preppy boy with a family fortune) as they first thought. He was self-funded through university. They said he wouldn't be at JCars if his company had been successful.

The "elite" is spoken about in terms of a person's access to private education and a matching lifestyle. They are seen to be more elite the longer they stay in private education (hence family background is key). That is combined with the elite status of the school or university. Miura-san, who has been working for 1½ years, is another graduate of a top private university in Tokyo. (Both Tabata and Miura's universities rank within the top five institutions in Japan.) He did not attend crammer schools to pass the entry examination; rather, he is very modest, and says he was lucky that he didn't need to go to a crammer because he lived in a district with a good public school. All his classmates were bright and went on to study at elite universities. His father is a senior figure in the national broadcasting company; based on his father's occupation, women said that Miura "comes from a good family" (いい家の子なんだよ). Miura himself did not identify with the elite because he was one of 1,300 students graduating from his department alone. Elite means wealth and minority numbers for Miura. His view is shared by some of the female workers as well.

In addition to class status, women's discourse regarding Tabata's fall from grace concerns masculinity in two respects: (1) his failed entrepreneurial project was seen as a vulnerability, which in turn was seen as a lack of masculinity; and (2) his career path inadvertently critiqued the traditional hegemonic ideal, which some women clearly support, which underscored the fact that a salary man's life is second best. Contingent facts thus also inform status evaluation. In Gordon Mathews' (2003) interesting work on men's *ikigai* (what is most meaningful and worthwhile in life) and new forms of masculinity, among a minority of younger men, entrepreneurship as "self-fulfilment" is an alternative to the salary man ideal of "commitment to company" (p. 113). Mathews shows that nevertheless many men continue to conform to the corporate ideal, if not for cultural then for institutional reasons.

Thus, when we incorporate this wider understanding of alternative masculinities (Connell and Messerschmidt 2005; Cornwall and Lindisfarne 1994; Nye 2005) in our analysis, we can better understand why Tabata's concessionary fallback into corporate life brought down his status among some women. From this example, we can observe that the position of various masculinities in the hierarchy is fluid, and women's evaluations shift between masculinities in a volatile manner. At times the salary man model ranks higher than the entrepreneur model and vice versa, depending on the situation and reading of other factors such as class. (We will come across other types of masculinity in chapter 6.) Equally, we can say that such women are pursuing ideals, and not evaluating the person fairly in his own terms. Reacting to this view, however, in Tabata's defense, some women said that the outcome was less important than his efforts. This applied both to his educational attainment and to career path.

Noting Tabata's career and others in their twenties and thirties whose jobs at JCars are their second placement brings us to the issue of mobility. For most Japanese men, as Tom Gill (2003: 145) explains in the introduction to his study of day laborers, "...masculine mobility is an escapist fantasy; ... imposed immobility [role as pillar supporting the household] and involuntary mobility [transfers] are facts of life for most men in regular employment, whatever the colour of their collar." Gill's day laborers are the antithesis of salarymen; they have freedom and autonomy (pp. 146–7).

At JCars then we witness a variant to this model of salary man masculinity: voluntary mobility as reality. Within JCars, some men, as do a few women, request and are granted domestic and more often international transfers. In addition, men talk positively about their interfirm experience, as taking pride in *shakai benkyō* (learning about society through experience, a diffracting of perspective). In this talk of knowledge, experience and mobility, the shape of individual narratives is constituted by and within social and cultural reality.

While only a minority constitute the voluntarily mobile group, excelling in wider work experience at other companies before settling at one particular company is a matter of gaining individual agency. This trend runs parallel to the greater cultural acceptance of mobility within elite labor markets (traditional, large companies and in the IT sector).[9] In this tier, social stigma is no longer attached to changing employment, whereas once it was seen to mark out the "commonly unsuccessful" (Valentine 1990: 43) or as a negative personal trait: maladaptive to work environments and ill equipped to fit in with hierarchical relationships. Changing employment is made practical by the "termination" of the lifetime employment system, an inevitable outcome of structural change in the economy and labor markets.

Young men like Miura, whose job at JCars is his first, incorporate this ideal of inter- and intra-firm mobility when reflecting on his career. He says "it's not good to stay in the same company, it's complacent, it prevents personal growth ... Also, one day, I'd like to experience work in an overseas office" (同じ会社にずっといるの良くないよなぁ。なんか自己満足でよくない。自分が成長しないよ。いつか海外転勤も経験してみたい。). Miura's statement appears to acknowledge mobility as a feature of late modernity, where a structural change from hierarchies to looser networks leads to greater flexibility, but, as Richard Sennett (1998) has noted in connection to the United States, this flexibility has had negative effects of job insecurity and declining loyalty felt by workers toward their companies. Returning to Gill's comment, we can see in this example of JCars, mobility as both aspiration/fantasy and reality. Nonetheless, in relation to Tabata, we might wish to question the assumption of voluntary mobility as a straightforwardly positive venture, given that a voluntary change of firm could have been initiated by a negative push factor (i.e., it is not entirely freely chosen for material or psychological reasons). As with many things, it is likely to be a result of a diverse combination of both negative and positive factors. This shifting terrain of multiple masculinities among salary men and of men in wider Japanese society will be of continuing interest to us.

Personality and Appearance of New Recruits

Thus far, we examined the selective criteria for entry into the company and for placement on the career tracking system as well as lived careers. We saw that status is negotiated through multiple factors, such as route of entry, allocation on the tracking system, through work content, class, and discourses around careers. We gained distance from purely structural accounts by taking note of practice in our view of status. This is a complex process: status takes account of both structural (objective) and experiential (subjective) criteria, and is highly situational. Thus, we have been moving through the discussion of social structure to cultural expectations and, in this section we examine status negotiation in light of discourses on personality and appearance—beauty, femininity, masculinity. This not only enlivens our understanding of gender and status in workplaces but also is vital to it.[10]

Personality and appearance affects the chances of success for both men and women in workplaces (McDowell 1997). To examine this in context of the recruitment process we start by expanding on Thomas Rohlen's (1974: 70) observations of how to evaluate a potential employee at a Tokyo bank: "[i]s he or she personable? Too forward, or too shy? Healthy? Attractive? Serious?" I take up Hori-san's entry experience to illustrate the personnel department's selection process from both sides, as a new recruit and as a junior member of the recruitment section who has one year's work experience. Her insights also convey a general knowledge of the department's procedures in the past.

JCars chose women who do not express strong opinions: 意見や疑問を持ち、争うぐらいまで言う人は採らない (Opinionated and inquisitive people who are willing to argue a point are not selected); and, 独創性の強い人は欲しがらない (Candidates with a strong sense of originality [personality] tend not to be wanted). This preference extends equally to male candidates, as the company prefers workers who follow the orders of their superiors. Therefore, JCars selects female and male candidates who can be easily molded to fit the company profile.[11] This is because JCars values tradition and hierarchy (although, not all large companies are similar in outlook, and there is further variation across sectors). Furthermore, constraints applied to one's body, through management of appearance according to expectations and anticipated responses, indicate the individual's willingness to identify with the group and to abide by rules. In other words, it is a show of respect for the social situation (Goffman 1963). Thus, as information is embodied, recruiters are more likely to select the candidate who appears to demonstrate the qualities of conformity.

Igarashi-san is Hori's senior in the same group by five years. Igarashi is involved in the first round of interviewing together with two other male colleagues. When interviewing female candidates, the personnel department judges women's personalities (*hito-gara*). Igarashi is aware that female candidates concentrate on putting across a favorable impression. In a 30-minute interview, she could distinguish between true and false statements. However, assessing personalities is difficult because the female interview candidates appear alike, all "make vague and noncommittal statements" (当り障りのないことを言う).[12] They "hide their personalities" and "feign innocence" (みんな猫をかぶる) and present a "pre-manufactured, made, or created self" (自分をつくってしまう). Male recruits are more expressive, and as a result, this makes for more interesting and entertaining interviews.

At the same time, Igarashi is aware of the paradox facing women in interview situations given the importance placed on gendered roles and cultural ideals of femininity within the organization. Kimura-*jōmu*, a board member of the overseas department, told me (somewhat controversially) how he thought "all Japanese girls are cute. Although they don't take work seriously. There are no girls who you would confuse for men as there are in the West. You just want to hug them all" (日本の女の子はかわいい。 仕事を真剣にはとらないがなー。外国みたいに男か女なのか見分けつかない子はいない。 みんなキュッとしたくなる様な子ばかり。). Muro-*kachō*, a section head in the overseas department, discussing his relationship with his teenage daughter said, "even with my own child, more than academic background, if she is obedient and sweet, that gives me sufficient happiness as a father" (自分の子供でも学歴より、素直でかわいければ父親として十分嬉しいねんで。). Hori said, if women "express their opinions forcefully" (はっきり言いすぎ), men criticize them for being "outspoken" (でしゃばっている); however, women criticize each other in this way as well, and fearless women become "frightening" (怖い) to others.

During the recruitment interview, the team assesses women's general appearance such as hairstyle and color, application of make-up, and the length of nails.[13] Male recruits are exempt from these criteria but others apply. Attention is paid to short hair, clean-shavenness, and tidiness.[14] Male candidates tend to have similar self-presentation to each other, whereas women show greater variation in appearance.

Despite the attention given to appearance, it is difficult to know the degree to which appearances affect the success of female applicants. This is partly because most women are recruited through personal connections. Nevertheless, women's self-presentation must respond to the recruiter's criteria. For example, Hori, who is not a *kone* recruit, prepared for her interview by dying her bleached hair back to her original color (black), removing her dark nail polish, and choosing a subtle color of lipstick. In presenting a conservative appearance she was "forsaking herself" (自分を捨てる). Hori sought friend's advice on self-presentation in interview situations. She also read women's fashion magazines and their special features during the recruitment season. They contain special articles on dress and make-up styles.[15] She wore a grey jacket with a skirt, although she "wanted to wear a trouser suit but [she] didn't want to stand out unusually, so [she] wore a skirt as it was a safe option" (パンツスーツにしたかったけど変に目立つのが嫌で無難なスカートにした。).

Hori's feelings give some perspective to how gender-specific values and expectations constrain female workers. It also reflects the power of patriarchal authority to shape women's bodies, if they wish to belong to a reputable organization like JCars. She finds the company dull and conservative, because the management consistently emphasizes formalities and regulates behavior and appearance. Hori begrudges having to conform to their codes. Her style is not conservative or classic. As a student she regarded her appearance as a statement of individuality and agency: by dying her hair, she rejected parental authority and conservative middle class conventions. At the interview with her hair colored black, she was submitting to parental, class, and managerial conventions based on a narrow if historically rich code of practice. Indeed, ". . . bodily control is an expression of social control" (Douglas 1996: 74), thus, Hori's comment about feeling as though she was forsaking herself at her interview makes sense. Moreover, her judgment to forego trousers for the interview anticipates issues of authority/power and gender in organizations. As dress has social significance, wearing trousers can have serious implications for women, as organizations are not gender-neutral and often position women disadvantageously to men (Halford and Leonard 2001). In rebuttal to this corporeal submission to conservative standards of appearance and behavior, she often raised the issue of wanting to leave her job. Furthermore, we might also note the certain irony in the tension between the recruit responding to the recruiter's criteria in this way and the recruiter's perceptions of the potential recruit, in that, as discussed earlier, new recruits are seen as hiding their personalities and presenting a pre-manufactured self.

The knowledge shown by the interviewee concerning correct attire and behavior exemplify the extent to which the company's values converge with those accepted in the culture at large. In childhood Japanese women are inculcated with the knowledge of desirable feminine attributes (Lebra 1984) that they will embody. Girls (late childhood to early adulthood) are disciplined (*shitsuke*) under various "codes" of femininity, such as "modesty," "code of silence," "tidiness," "courtesy," and "elegance." At the same time, girls undergo "moral training" and acquire attitudes and behavior of "compliance," "diligence," and "endurance" (Lebra 1984: 42–6). These attributes fit the Confucian ideology of good wife, and good mother—*ryōsai-kenbo* (Sievers 1983; Uno 1993), where women follow their fathers, then husbands, and finally sons (Iwao 1993; Kondo 1990a). Parents insist on immaculate manners from their female children but not of males. While eating, the distance of the elbows to the side of the torso cannot be too far or too close, chopsticks must be held correctly, and one's chopsticks must not roam indecisively above individual dishes (*mayoi-bashi*), bowls and plates must be received with both hands.

The recruiters at JCars look for feminine attributes in their new recruits. This high degree of interdependence between parental and company values mean future workers are more likely to abide by the dominant management discourse. Moreover, the management discourse legitimates and naturalizes its position through its association with conventional ideals in wider Japanese society. The media also support its position. Articles appear in popular women's magazines, newspapers, and on television, often with titles such as "how to become a successful career woman," proliferate and reinforce the conventions associated with corporate dress and appearance, and by so doing, the dominant discourse of management seems to succeed in cross-cutting generational boundaries; tradition sticks and the new cannot be incorporated. Tanaka (1990, 1994) found in her study of Japanese advertisements targeting women how the use of words such as "intelligence," "individualistic," and "feminism" was associated with different meanings to their dictionary definitions. The particular light-hearted use of these words as fashionable terms served to reproduce stereotypical images of women and reinforce existing patriarchal social values where a woman's place is understood to be in the home (ibid.).

Corporate values and the expectations it places on recruits not only constrain female workers, it also shapes their male peers. Indeed, the dominant discourse concerning men's appearance defines masculinity in a corporate style that is closely tied to the notion of the devoted worker (for detailed analyses see Dasgupta 2000 and Mathews 2003). Men face serious consequences if they refuse to abide by the code, but this is not equally true of all departments. Elite departments observe strict rules compared to other departments. Members of such departments are explicitly and implicitly called upon to represent the company (more so than other members of the workforce) and therefore must dress the part. Individuality is unimportant. Rules of ideal appearance are strictly enforced. Men should not wear colors other than white as they distract workers, in the same way excessive talking might. In the personnel department, which sets the dress code regulations, colored shirts are seen to disrupt the homogeneity of the workforce.

In practice, one young man was demoted from the personnel department for wearing pink shirts or rather for his disobedience as he was repeatedly requested not to wear pink shirts by the head of department. Female workers can be criticized by managers for wearing their skirts too short. Limits and proprieties are set for both men and women, but generally, men face serious career-sabotaging consequences for noncompliance whereas women do not. What defines proper appearance for females often, however, filters into workplace politics and is used in the contestation of status between female workers. Management uses appearances to judge female workers because women are judged for being women.

There is a correlation between cultural and corporate values: both place importance on the role of women as wives and mothers, and value feminine appearance and behavior. Female colleagues use appearance as an important means of evaluating fellow female colleagues because appearance is one criterion through which female workers are judged against each other by those in management positions. It is a circular argument. The importance women attach to heeding the management discourse is obvious. It is because as workers, they are in direct competition with each other in terms of salary, bonuses, performance evaluations, and promotion. Of course, women are also in competition with men, but, as discussed previously, as a structurally subordinate group within the male-dominated organization, competition between women entails significant purpose. Moreover, the ambiguous nature of the link between objective and subjective criteria that enters into the assessment of an individual's performance at work induces appearance to take on an ever more pertinent significance.

Male workers have restricted modes of expression via dress, likewise a woman's individuality is ironed out by standard issue uniforms.[16] Nevertheless, women find creative ways of expressing their individuality, beauty, and femininity. Some women's efforts to alter their uniforms are linked to their hopes of influencing the male gaze. Skirt hems are shortened or widths narrowed to accentuate the shape of the hip. One woman, despite her discomfort, wore her waistcoat in a smaller size to accentuate her bust and narrow waist (*kubire*). Appearing attractive to male workers is crucial; in particular (and implicitly toward) those who influence women's careers or (explicitly to those involving) romantic relations. But women do not invest in beauty merely in response to the male gaze; rather, they do it for themselves as a productive activity in self-expression. As Judith Butler has shown, gender is a "constituted *social temporality*":

> ... [G]ender is an identity tenuously constituted in time, instituted in an exterior space through a *stylized repetition of acts*. The effect of gender is produced through the stylization of the body and, hence, must be understood as the mundane way in which bodily gestures, movements, and styles of various kinds constitute the illusion of an abiding gendered self. (Butler 2007: 191, italics in original)

Femininity and beauty is "performed," as from the quote above, through elegant mannerisms, hold of the body, including the hands; proportion;

hairstyle and color, by whitening of skin; and by clothes and accessories worn to and from work. Beauty and femininity: a gentle manner of speaking, soft laughter, not the explosive kind that erupts spontaneously from the belly in reaction to a good joke. Beauty and femininity: a body groomed to perfection. Make-up is always worn, and carried in lovely pouches. Some women never leave the house without "base-make"—foundation, eye shadow, mascara, and lipstick. The nails are painted, even if they are in a clear polish. Nail art is popular at present. If nails are too long it is considered *fuketsu* (filthy). Legs are never bare, even in summer, and flesh-tone sheer stockings are worn, even with peep-toe and strappy sandals. Surfaces that come into direct contact with outside dirt are shielded. Jewelery must be delicate and understated (*sarigenaku*), tasteful, perhaps with diamonds. Hair must not be left black in its original state, a chestnut hue is good, but not too light. The ideal texture is straight and glossy with each strand of hair being thin (as opposed to coarse), soft to the touch, and lissome (*shinayaka*)—the hair too takes on the sexiness and limberness of the youthful female body. Unbeknownst to the individual, one woman was called *purin* (crème caramel), because her hair was too light and as it grew out her dark roots showed in drastic contrast. The management, however, prefers black hair. It must be styled neatly but not too sexily. Similarly, lip gloss should not be too glossy or wet looking. A direct reference to sex should be avoided.

Beauty and femininity: thinness. Women eat slowly and delicately, calculating calories in small mouthfuls. Women eat their minute portions very deliberately; men can eat as though they're hungry. Some women only eat diet supplements (*sapuri*) in the form of crackers, cookies, or *konnyaku* jelly in peach, grape, and grapefruit varieties (*konnyaku*: fibrous, zero-calorie paste made from devil's tongue, a root vegetable; known for its detoxifying properties). One day for lunch, I bought three bread rolls from a *depa-chika* (underground food hall in department stores) bakery. All the women in the group roared with disapproval. "You're eating three?! Kuri-rin [author], you're such a failure. Normally, you only eat two!" (三つも食べるん？！栗りんあかんなー。普通二つやで!). Mortified, again, on another occasion when my convenience store–bought *bentō* raised alarm; apparently, it was man-size, although it was not labeled as such. One doesn't carry around two liter Evian bottles (this applies equally to men) as women do in London. The drinks must be small, like the body. To perform the feminine one conforms to portion control.

Thinness is the cultural standard. Araki-san, who learns and teaches dance, was told to lose weight by an ex-boyfriend, a medical doctor. He said if she got down to 48kg, he would buy her designer clothes. She is 170cm tall. Her current boyfriend says that her stomach is too big. It is not. Dialectic between women's aspirations and men's desire is based on the impossible; the performer and audience are both at one with each other in upholding a "mode of belief" (Butler 2007: 192).

Beauty and femininity: the latest beauty products and techniques are vital, for example: specialized products bought by mail order such as a rare and rather mysterious potion made from equine extract used to promote eyelash growth; debates on eyelash perms and detailed methods of hair depilation; getting a Western-shaped eye with the doublefold through mini cosmetic surgery (*puchi-seikei*). Some women attend aesthetic salons (*esutè*) on a weekly basis and have all-body treatments to lose fat and gain a shapely, sexy waistline and legs.[17] The sexy waist, mentioned above, symbolizes Japanese femininity; the place on the body where eyes focus. Since the 1990s, however, more attention is paid to the beautiful cleavage and desire for large breasts. Teeth straightening and whitening is now popular among celebrities or *tarento*, the trend slowly trickling down to the masses.

Beauty and femininity: sartorial elegance for commuting to work. Never should one wear loud or casual clothes, particularly trainers. Hori once declined an invitation to an afterwork drink because she had worn her canvas Converse All-Stars. Women check each other's outfits in the changing rooms, at times looking and commenting deliberately to express disdain or approval. Naturally beautiful and feminine: senior women's appearance is conservative, rarely do they dye their hair brown or wear tight uniforms; tasteful and not showy. They appreciate natural beauty and do not take part in performances of the kind described above.

Beautiful and feminine: *kirei. Bijin.* The cultural standard. Li-san, who is in her fifth year at JCars, says "in Japan everyone says of everyone else: *kirei* (beautiful) or *bijin* (beautiful woman). When I cut my hair the women at work all chime '*kireeeei!*' In China only *bijin* are called *bijin.*"

What women gain is aesthetic and moral advantage, a knowledge of one's value, or body capital (Bourdieu 1984: 202–7). Most important, appearance is noted, memorized and used in conversations among women that set one off against the other and is linked tenuously to a woman's status, personhood, popularity, and evaluation of competence on the job (cf., Traweek 1988).

Clearly, status is defined in many ways. In the language of anthropology some of these differences—economic, cultural, and body capital—indicate class differences. However, in Japan, differences are not articulated in terms of class, translated as *kaikyū* or *kaisō* (Atsumi 1980; Kondo 1990a: 314, n. 20; Nakane 1970; Sugimoto 1997). Rather, as we have seen, differences are talked about in terms of the attributes themselves (based on possession of economic, cultural, and body capital). Current terms used to describe socio-structural change, in particular the growth in social disparity, are *kakusa* (social inequality) and *karyū-shakai* (downward drift through social strata: people once identified with the middle stratum sliding toward the lower middle and lower classes).[18]

Female Office Workers, Meaning, and Discourses

The remainder of this chapter deals with the status of female office workers as a category in society (we consider men in chapter 6). Thereby my argument comes full circle from where it began at the start of chapter 2. We examine how my subjects made sense of their objectification and positioning as "office ladies" or "OL" in the light of our earlier discussion on status negotiation. Their experience shows that social representations are alienating in their fictitiousness. Equally, their comments testify to the transformations taking place among women concerning the meaning of work in their lives, and this is grounded in changing workplace culture. I have confined what is given here to what emerged in conversation. The analysis of how OLs are represented in popular cultural writing and visual images requires a larger project that I do not pursue in this book.

The spirit of this section follows the approaches of Denise Riley's (1988) study of the discursive historical formation of "women" as a category and her point about temporality that collectively "women" are an instable category, and, although we are women, individually speaking that is not all we are all of the time, and we might reject, adopt, or hesitate in relation to how we are described; Rosemary Pringle's (1994) tracing of dominant discourses that define British secretaries to explore the production of a variety of meanings; and, Konno Minako's (2000) socio-historical analysis that exposes the origins of gender segregation in Japanese workplaces and charts the changing meaning of gender categories in relation to office ladies. Deconstructing the categorizations of women is a universal task.

Glamor

As many have pointed out, the term OL is defined through the use of loan words (*gairai-go*). Loan words designate foreign concepts that have no prior equivalent in Japan; some entail connotations of glamor that reflect a tendency to admire the Occident. Glamor entices young women who, after junior college or university, enter into the world of work; the image evokes liberation from the education system and the cultivation of maturity and independence. But many are not released from the home.

Some of my younger subjects who had been at the company for a year described the perfect office lady as a clever woman, a graduate of a prestigious university; coming from a good or wealthy home (the two being linked); allocated to an elite department within the company; confident and firm with colleagues when need be; can conduct herself well at all times and particularly under pressure; has determination to finish all jobs;

pretty; dressed elegantly; wears desirable, fashionable, branded goods from accessories to clothes. The younger women aspire to be like their favorite role models among older female workers. They speak about them in adoration as they do of the men they fancy. Inspiration is also drawn from actresses on television. Some junior women label the contemporary image of the OL as cool (*kakkō-ii* in contrast to *kawaii* or cute). This word is typically used to evaluate the desirability of men among women, whereas men rarely, if ever, assess a potential partner through this description, although, they refer to some of their career-minded female colleagues in this way. The use of *kakkō-ii* in a context of women describing other women suggests that OLs are catching up with male colleagues. In other words, the term connotes the celebratory sense of successful women competing in a male-dominated environment. However, women who are seen to have acquired masculine traits, "man-like" (*otoko mitai*), are criticized. As we saw previously, many think that a woman should never neglect her femininity, and loss of femininity is interpreted as a shameful episode of role confusion.

Independence

The notion of independence also forms part of the image of office ladies, and this is as relevant as physical factors to do with appearance. The cultural expectation of an OL is conveyed by the phrase 社会人になれば自立しなければならない (the need to gain emotional and economic independence once one becomes a member of society).[19] This is typically cited when subjects are asked to define a single feature distinguishing the status of student from worker. *Jiritsu* (independence) in this context has both psychological and socio-economic meanings.

Jiritsu in the social sense is about knowing one's place in relation to the other. As a worker this means relearning the practical application of polite and honorific forms of address that they acquired initially within hierarchical relations at school and university. Multiple forms of honorific language are called into use in workplaces. For example, in distinguishing between in-house senior managers and managers in external companies, and between the various ranks of managers inside JCars. Furthermore, women address men in a more polite form than they would take among themselves.

In the economic sense, *jiritsu* refers exclusively to earning an income, and does not necessarily take on board financial responsibilities such as paying rent, managing a mortgage, or the personal domestic responsibilities such as cooking or laundry. This is because most working women live with their parents until marriage where rent and meals are provided for. Most women contribute a set amount to their parents' monthly food budget (between ¥30,000 and 50,000), or helped with the household chores in recompense. Some kind mothers appreciate their daughters' gesture and place the full amount into their daughters' savings account.

Negative judgment against a woman living independently of their parents before marriage does not allow women to experience full independence. Company policy is in part responsible for this as many companies do not recruit individuals who cannot commute from their parents' home. Japanese early womanhood and working lives are atypical of the Western experience. There were exceptions to the general rule of course. Ono-san lives in an apartment owned by her parents but they live abroad; she acquires legitimacy given her father's status as a JCar employee. Sakamoto-*shunin*, who exudes self-confidence is one of the two female employees to have held overseas posts. She has her own apartment, and defies workplace conventions and the sanction of negative opinion of others. During one lunch outing together, she said that she wanted her own space, away from her parents, and to be self-sufficient. It means more to her than a practical solution to a long commute.

Career Woman or Office Worker

The personnel department was among the first departments to rid the label *hosa* (supporting roles) from women's official job descriptions. As we read earlier, male managers say that women are given equal job content and responsibility to junior men in the office. They stress that women's efforts are valued, and sympathize with women for the cultural barriers impeding women's career paths. This managerial talk exists awkwardly in relation to actual managers who maintain the structural causes of gender segregation women experience, for instance, in terms of promotion to the career track. This shows the many levels that exist within what is taken to be the dominant discourse of management.

Some women are critical of the label "OL" and avoid this self-description. This demonstrates a transformation in women's experience of status. It testifies to the everyday processes of status negotiation. Ando-san, a career track worker in her sixth year of employment in the overseas department, considers the term derogatory, undignified, and inconsistent with what her job involves. She says "OL" reflects the wider ignorant social perception based on outdated images of women in supporting and secretarial roles, which is inapplicable to "[career] women who work incessantly" (仕事をバリバリしている女の人).

In Ando's opinion, office ladies and career women should be ranked separately as two distinct categories of female workers, because not every office lady is a career woman; while every career woman is essentially seen as an office lady. Technically "career woman" describes a woman with specialized knowledge or in a managerial position (*buchō* class or higher). Ando "feels aversion/negative pressure" (抵抗を感じる) toward the word OL. For example, when filling in occupational categories in forms she chooses the gender-neutral term "office worker" (*kaisha-in*), thereby avoiding the gendered conception of work intrinsic to "OL." As though in a linguistic fury capable of erasure, Ando says "the word office lady is a Japanese invention constructed using English words, the combination does not exist in standard English" (OLの言葉は和製英語だから英語には存在しない). She highlights the illegitimacy of the label. It is all but an objectifying and homogenizing cultural construct that takes its meaning from an earlier referent, women as "office flowers" (*shokuba no hana*).

Conversely, to the younger generation of workers "OL" connotes glamorous images in terms of an expensive lifestyle (of traveling, going out often, shopping for designer brands) and status (well educated and competent job wise). Yet "OL" is undesirable for older women insofar as images of an attractive worker performing simple menial tasks still remains the prominent one from which they say some older generations of managers and wider society view women's roles in the workplace. If a gendered terminology is unavoidable, such women prefer "career woman" or "*kaisha-in*" to "OL." The meaning of "OL" thus carries different connotations among women. The term continues to take on new meanings. It is not that younger women are indifferent to being compartmentalized as female office workers. After all, work comprises a prominent part of the image. Rather I feel that the feelings evoked in relation to both meanings are indicative of the changing nature of the workplace and women's different position and role within it. Older workers have had to endure more changes and pressure of gender segregation, but, because they did, the norms have started to shift. Younger workers have become desensitized to a term that in their eyes has lost some of its pejorative connotations and in some instances has taken on a positive association. Even in the six–eight-year age gap between Ando and the youngest recruits, the meaning of OL has shifted; it continues to shift in relation to socio-cultural change.

Summary: Discourses of Status

As we have seen throughout this chapter, the issue of status in the workplace is characterized by what linguistic and cultural analyst Mikhail Bakhtin (1981) calls "heteroglossia."[20] Heteroglossia at JCars accommodates a multiplicity of subject positions and interpretations of status. These exist in the workplace at one time. Workers make sense of status positions in a language that is at times consistent with and at other times in contradiction to the dominant discourse. Simply put, how workers define their own status as well as that of others exemplifies a struggle between incorporating the dominant system of meanings into one's own way of interpreting status and attempting to break free from becoming subsumed by dominant values. In Bakhtin's terminology, these forces are expressed as "authoritative discourse" and "internally-persuasive discourse" respectively. Bakhtin's theory of language states that

> The word, directed toward its object, enters a dialogically agitated and tension-filled environment of alien words, value judgements and accents, weaves in and out of complex interrelationships, merges with some, recoils from others, intersects with yet a third group: and all this may crucially shape discourse, may leave a trace in all its semantic layers, may complicate its expression and influence its entire stylistic profile. (Bakhtin 1981: 276)

In summary, then, the various opinions of men and women as they relate to issues of status are situated within a framework dominated by an economic and ideologically loaded logic preferred by the management, and in this the need to achieve gender equality interacts in contradictory and complimentary ways with cultural, social, historical, and professional norms associated with the role of women. Thus, the management *attempts* to maintain a degree of sovereignty over other competing discourses thereby controlling the official definition of women's status. The management asserts economic justifications for women's structural positions and, by presenting this in the language of logic and reason, masks its gender-biased policies. This is evident when the company's recruitment methods for men and women are juxtaposed. The elaborate and costly methods taken for recruiting men, and by contrast, the recruitment of female university graduates based only on personal connections and the limited method of recruitment of female junior college graduates, not only reflect the company's cost saving regime, which is also evident in the ceiling imposed on women's wages and promotion, but these differences simultaneously establish, convey, and reinforce the management's view of female workers as being easily replaceable members in the workplace.

Cultural notions that tend to stress the importance of women as wives and mothers and the predominating concern with marriage that make women seek early retirement feed in to support the economic logic of the company, because not all women necessarily want long-term careers or to be assigned the same amount of work as men. Further, the company is reluctant to change the tracking system by which they maintain cost-efficiency, at the expense of achieving status equality between men and women, because the system also provides a means to legitimize the existence of this inequality without appearing discriminatory.

Despite this, it is abundantly clear that not all managers keep women in low status positions and many actively encourage women to convert from the clerical to the career track. By this it is shown that the management is not composed of individuals with similar interests and opinions. I do not discuss in depth reasons why such dissimilar interests coexist within the management as it seems sufficient to point out that the management, which appears unitary, is, in fact, heterogeneous. The "dominant class" can be misconstrued as being unified, yet, clearly, it is comprised of conflicting interests or "cultural practices" (Bourdieu 1984: 270). The management might appear unified in its objectives because, like language, "centripetal" forces are in operation, attempting to centralize and unify the alternative discourses internal to it (Bakhtin 1981: 270).

That said, despite the disadvantages they face, women negotiate their status, raising it through other avenues. Important is occupational role. Women experience changes to their occupational roles as a measure of greater equality. This is significant to the way we understand women's status in the workplace. But again, the significance of the modification to women's occupational roles is interpreted differently among women, where some women find their current roles to be sufficient and did not require this change in work content to be reflected necessarily officially, say, by promotion to career track positions. On the other hand, career-minded women tend to consider the company's treatment of women as unjust and want further changes to status, that is, for their work content to be remunerated in structural or objectively perceivable terms. The restraint of the former, in part, is due to the fear of emphasizing, though unintentionally, the inequality between women that originates from differences in educational status. Other women are openly competitive. While women's status can be conceptualized collectively, and women referred to as OL, doing so contributes to a vast generalization because individuals are predisposed to various ways of thinking about what the role of women should be.

In connection to status, notions of appearance are important to male and female workers and equally to the management as a means of assessing the ideal worker. As we will see in further detail in chapter 6, the dress code also applies to the assessment of men but exempts particular men, and, in this chapter I have shown by concentrating on women and men how the issue of appearance is tied to a larger discourse about cultural notions of femininity and masculinity that intersect with notions of class. As appearance is significant to the management discourse it also becomes a means through which women evaluate status among themselves. Appearance, thus, is a way women are judged in the workplace by managers, as much as it enables women to judge each other in terms of a gendered individual and as an ideal worker.

Yet, we also saw the tenuous link between appearance as a means of assessing status. In particular, career-minded women experience being defined as an "OL" to be superficial and demeaning because it is an image based largely on factors of appearance than on performance and ability. Whereas, younger workers did not find the label "OL" problematic, career women preferred the use of the gender-neutral term "*kaisha-in*" to describe themselves as members of the workforce. In this study of differing meanings held among older and younger women around the image of OLs, we saw that among younger women, work is central to the image of OLs even while glamorous aspects of appearance and lifestyle are also praised. We can interpret this as an underscoring of socio-cultural changes over time.

The status of women cannot only be dealt with through a consideration of wage and promotional structures, which do not account for the experience of occupational roles and changes taking place to the interpretation of the meaning of women's status currently, over time, and according to context. As status perceived purely through structural measures is inconsistent to that experienced in practice, some women view their status as being low in reference to wage and promotional structures, yet as equal to men when the focus shifts to a discussion of their experience of occupational roles.

Thus, status is not a given, but is constantly negotiated on an everyday basis. Status is made and remade through this intersection between structural premises for judging women's status and a more practice oriented perspective of assessing status. Further, status is complicated by various personal views sustained by each worker that is determined by the degree of adherence to socio-cultural norms, and by the company policy that takes advantage of these socio-cultural norms. Women have a high degree of awareness toward company policy, which simplify women's employment down to the category of gender. My point is that women are conscious-reflexive subjects who do not allow themselves to be treated in abstract terms, thus, women do not always allow gender to be the determining factor that dictates the experience of being a worker.

In the next chapter we consider the theme of community in greater depth by exploring how social relations are structured at JCars, analyzing how the new recruits familiarize themselves with work routines, senior colleagues including managers, and peer groups (i.e., vertical and horizontal social relations).

CHAPTER 5

Linguistic Spaces of Vertical and Horizontal Organization

So far in our analysis, we have seen how economic, legal, and socio-cultural forces structure employment practice, career trajectories, and occupations within JCars. We have begun to gain a nuanced understanding of the manifold ways in which status is negotiated on an everyday basis. The context for looking more closely at the experience of social relations within JCars in the preceding chapter was firm entry and the career tracking system. This chapter introduces another way that interpersonal relationships are structured, namely vertical and horizontal patterns of interrelationships operating among fellow workers. These function within the larger group units of the section, the department, then, the company. For the most part, individuals are expected to follow a prescribed way of relating to others within these relationships. The company's expectations parallel Nakane Chie's (1970) description of Japanese social organization. Thus, I re-examine Nakane's model, its features and limitations, in relation to my observations at JCars. As an effort to understand the psycho-social and linguistic space of individuals and interpersonal relations within vertical and horizontal organization, I contrast actual experience against expectations of how these relationships should function. This allows us to engage in a deeper analysis of the meaning of community; meanwhile, this sets the groundwork for subsequent chapters of this book that analyze the relation of experience to ideals (of gender, status politics, and ideology) in the community, albeit through other concepts such as ritual, time, and spatial practice.

JCar's Social Structure = Nakane's Model

In Nakane's (1970) model, hierarchical principles organize social relations into vertical relations that are experienced between senior (*senpai*) and junior (*kōhai*) workers, and horizontal relations that are experienced between individuals of the same rank or horizontal stratum (*dōryō*). Thus, vertical and horizontal relationships function by linking individuals with different qualities and those with the same qualities respectively (p. 24). As a vertically structured society these sets of relations become "...primary factors in relationships,...constitut[ing] the core of a group's structure" that bring a sense of "cohesion" and "stability" to the group (pp. 24–6). These ties between individuals are strong because (1) individuals belong to a particular group ("frame"); (2) membership in a group is the essential basis of Japanese social structure (pp. 1–8); (3) contact between individuals lays stress on the emotions and morals of belonging to the group (pp. 9–23) that breeds loyalty, dependency, and affection between individuals within vertical ties (p. 67). These characteristics of vertical organization enable it to function. It is in this sense that the operation of a strong consciousness of ranking order effectively exerts control over individual behavior and thought (pp. 24–38). But Nakane's interpretation of these vertical and horizontal relationships runs into difficulty.

Vertical and horizontal sets of relationships essentially infer something about the *quality* of interrelationships, yet an explanation at group and societal level is too broad to elucidate this as fact. Expansive evidence drawn from a comprehensive range of contexts shows how vertical relations function within different groups in Japan, from the historical to contemporary: large private companies, government factions, the military and religious groups, and so on, including cross-cultural comparisons with China and India. This demonstrates how the accumulation of individual sets of vertical interpersonal relationships sustains the harmony, stability, and continuity of groups and society. Yet these relationships themselves are not studied because, as Nakane admits, there is unpredictability and variation in individual behavior (p. 32).

Horizontal relations are also fundamental to the structure of groups, but do not receive much attention because they function less dynamically than vertical relations (p. 67), and the reality within the horizontal stratum is too complex that, without simplification, is impossible to account for as a group comprised of individuals of equal qualities (pp. 27–8). In fact, what the internal composition of *dōryō* ("same rank") group members demonstrate is a vertical structure differentiated by "age, year of entry or of graduation from school or college" (ibid.). In effect, this relationship is pseudo-horizontal. Nakane has opted to focus on the shared quality among individuals, that is, rank, which functions to unite individuals comprising the *dōryō* group, by setting aside the internal dynamics of the group produced by these differences between its members. These differences are passed over in order to let the *dōryō* group stand as the egalitarian counterpart to the vertical relations between senior and junior members of an organization.[1]

Interpersonal relationships are discussed in terms similar to a molecular equation showing the composition of a chemical compound (pp. 42–4).

But it is known that scientific truths are constructed facts, which convince people without them realizing they have been convinced (Latour and Woolgar 1979). Furthermore, with a structural-functionalist method one is hardpressed to describe anything other than ideals. Reality is something different. The fundamental difficulty of Nakane's model parallels the problem with colonial and contemporary anthropology, whereby the "ideological conception of social structure and culture" disguises the fact that the "formation of different forms of discourse come to be materially produced and maintained as authoritative systems" (Asad 1979: 619–24). Models of hierarchy become authoritative systems because they capture, if only by essence, an important aspect of social organization, which become useful for the anthropologist to summarize the actual experience of hierarchy (Appadurai 1988: 45–6). What is needed, therefore, is an understanding of the intricacies of interaction between individuals on a one-to-one basis.

At JCars the terms *senpai*, *kōhai*, and *dōki* (same year of entry), rather than *dōryō* (same rank), are used to describe vertical and horizontal forms of social relations. The *dōki* group refers to individuals of the same year of entry and corresponds to the way my subjects understand the horizontal relationship. An ethnography of a large Japanese white-collar office offers a similar view:

> The year a person entered the firm was considered important, and this importance was reflected in people's ongoing concern to identify one's *dōki* (those who entered the company in the same year). The word was used frequently by both men and women to describe their relationships to one another. Anyone who was not one's *dōki* became either one's *senpai* (one's senior) or *kōhai* (one's junior), valued terms not only in the bank but also in Japanese society in general. (Ogasawara 1998: 48)

The *dōki* relation describes workers of a single year group, rather than incorporating the entire range of individuals of the same rank whose year of entry can vary between two and nine years. Thus, *dōki* narrows down the horizontal group of individuals which Nakane (1970) had labeled *dōryō*. At JCars, the category *dōki* group, rather than *dōryō*, is more often relevant to the workers' experience of the office and their interrelationships. Yet, individuals within the *dōki* group vary according to the month of entry during the year (the majority of recruits started work in April); age; type or length of tertiary education; and type of worker—temporary or full-time. Further, other ways of assessing people as discussed in previous chapters contributed to the production of differentiation within the *dōki* group.

On observing vertical and horizontal relationships among my subjects, although these relationships occur between men and women, it is clear that the most frequently occurring pattern of interaction, and by this perhaps the most significant in terms of its importance to the workers' everyday lives, is between same-sex members. Therefore, our view of the workplace is not so much split down gender lines of men versus women, but more subtle, interrelating, and complex. Thus, it would seem vital to explore women interrelating with men *and* with other women within vertical and horizontal relationships. We build on the work of sociologist Yuko Ogasawara's (1998) account of gender relations in a Tokyo bank. This is a perceptive account, however, the scope of her enquiry was limited by her critique of power via the framework of resistance, which necessarily views women as pitted against men. In Ogasawara's account, interrelations among women are notable in the inability to achieve solidarity, for example, through differences in age, education, tenure, pressure to marry, and retire (pp. 44–69). Conversely, in the preceding chapter, we saw how women in the personnel department at JCars display discursive solidarity based on occupational roles. This leads us to take a different focus. Analyzing female interrelationships in their own right (not in terms of whether they can form organized protest) can add to our knowledge of organizations and workplace relationships, particularly in this time when the nature of work and occupations is changing. Moreover, as argued in this chapter and elsewhere, social relations in the firm are predisposed to disorder (*midare*) and differentiation, and from this state, management and workers try to shape order and their ideal community.

Vertical Relations: The Tutor System

A seminar for new recruits (*shinnyūsei-dōnyū-kenshū*) is held in April, where seven members of the personnel department organize a series of detailed lectures dealing with the various aspects of working at the company, that is, information about company operations, policy, products, morals, and rules of conduct appropriate for employees. All new workers attend this week-long seminar that is held at one of the company facilities. They are being groomed to become workers. The important aim is to spend time getting to know one another. After this seminar, while male recruits go on to complete further training in subsidiaries across Japan for a further three months, female recruits, exempt from this additional training, go on to the office or research center to which they have been assigned. As we have seen previously, the women know which track they are allocated to, and, once at the office, they are given notice of their placements in different departments.

Job Supervision

In the department new male and female recruits are assigned to a senior worker in a tutor role who takes them through the various aspects of the job (*gyōmu-shidō-in*). The new recruit is given a small task related to a larger project coordinated by their section or departmental manager. However, managers of different levels of seniority within one department can give a recruit any number of tasks. Thus, as the new recruit can be placed within the supervision of many senior workers, a *gyōmu-shidō-in* is appointed to each new employee; their role is to act as a point of reference for the new recruit who learns to juggle many tasks and relationships. The newcomer is tutored in practical skills relating to organizing and goal-setting, proper attitudes toward work, as well as understanding the methods and details of their tasks. Male recruits are only tutored with respect to the job itself, whereas female recruits receive additional instruction in managing the body and the feelings of others, and the atmosphere of the workplace. Arlie Hochschild (1983) describes this kind of work as emotional labor. But emotional labor is performed by both men and women as we will see below.

A brief "manual of the tutor system" (*shidō-in-seido no tebiki*) compiled by the personnel department is given to prospective tutors (male and female) in mid-March. The manual stipulates the purpose of the tutoring system: "to tutor/guide the employee in order to enable them to take an active part as a wholesome member of society" (健全な社会人として活躍できるよう指導すること) and "as a member of the workplace" (職場の一員として). Prospective tutors are told, "you are a model [for the new recruit]" (お手本はあなたです) and should prepare for this role by setting an example of oneself as a good worker. As tutors they are asked to "revise and relearn [if necessary] the [details of their] job" that they will teach the new recruit (仕事の見直しをしてください). Tutors are asked to "prepare a tutorial plan in detail" (指導計画を綿密にたてましょう). Further, they are asked to "work hard in the creation of a warm workplace atmosphere or mood" (暖かい職場のムード作りに励んでください).

As part of the tutor system new female recruits keep a work diary. The diary is kept daily for the first two weeks since starting work, and on every Friday thereafter until the end of May, and at the end of every month for the remaining four months between June and September. New recruits submit this diary regularly to the "sister" and work tutor (male). The dairy is then circulated in order of seniority, to the section head, the departmental head, and to a board member who give written comments of advice and encouragement. The personnel department envisages this diary method as an efficient way for senior workers to keep a close eye on the feelings of the recruit as she familiarizes herself to the new work environment. Moreover, the diary is considered by management to provide a context of communication for the uneasy recruit, who might find it helpful to express difficulties and address questions to senior individuals. This indirect and less immediate method of communication is thought to be less intimidating for the newcomer. In the vertical relationship between seniors and juniors this diary is a way to "deepen communication and understanding" (コミュニケーションと理解を深める). Following Hochschild (1983), we see here the importance of managing feelings to maintain the atmosphere or emotional state at section and departmental level. In this example, the communication and monitoring of feelings precipitating from work-related tasks themselves are managed at multiple levels, and the management of feelings of the new recruit is the responsibility of the entire organization, including the worktutors and the ranks of senior men.

Most women in the overseas and personnel departments gave me access to their work diaries. Broadly speaking, when women are uncertain about the work procedure they ask the senior workers directly involved in their day-to-day work. Thus, the diary is an important source of contact and feedback from workers of more senior rank, to whom they would not normally go for advice (e.g., board member). Up to this point, the diary is functioning as expected. However, the generic advice and comments they receive do not equate to a furthering of communication and understanding. For both the new recruits and senior individuals who work on the same project, the notion of furthering communication and understanding is somewhat paradoxical, as the distance between them in terms of status means that they do not communicate at that level required to achieve it. Although the diary is designed to facilitate communication within vertical relations, the women come to see the diary as a formal exercise that, counter to its official purpose, emphasizes the distance between the new female recruit and older male workers within these relationships. This highlights the difference between company ideals and the experience of new recruits both in terms of the meaning of communication itself and personal expectations envisaged with regard to being understood via communication.

Sister Tutelage

Female recruits are assigned to an additional senior worker (also female), a sister (*shisuta*), who tutors them in the general aspects of working in a business environment. The sister tutelage system (*shisuta-seido*) also aims to *create* a warm working environment, in a similar way as stated above in the guidebook for prospective tutors. The *shisuta* is responsible for offering emotional support (*sōdan-aite*). She will teach new recruits about time keeping (*jikan o mamoru*), the rules of formal greetings (*aisatsu*), the appropriate use of polite forms of language, correct telephone manners, and message taking. She also explains the organization of the workplace: arrangement, cleanliness, and cleaning (*seiri, seiton, seiketsu, seisō*). The sister also shows the junior worker the location and management of office stationery and use of the photocopier and other technologies used in the office.

The performance of these behaviors that falls on women exhibit a kind of caring for others' feelings in the workplace. This is why such tasks and behaviors are valued. Ideally, these are accompanied by corresponding feelings that denote concern and care. This does show the extent to which job roles are gendered, but it is also clear that the care and attention to feelings go beyond the question of gender because, as we saw above, both men and women must acknowledge and demonstrate the importance given to the display and modification of feelings. It is fundamental to the atmosphere, image, and rules of the company. At a collective level, this "sister" arrangement appears to reinforce images of sexual segregation from the perspective that women are only granted autonomy in a separate social sphere to men; this is an ingrained cultural feature of Japanese society (Ueno 1987a). However, taking a relational approach to gender will enable us to see how the sister relation unfolds over time and across various situations in relation to other relationships that include male workers.

It is interesting that JCars marks this interrelationship between women with a kin-based role loaded with responsibility. This is culturally fitting as sibling relations are conceived in such terms. Liza Dalby (2000) shows how in Japan, the sisterly relation indicates hierarchy, a position of older and younger, an unequal relationship, which is counter-intuitive to a Western notion that assumes a lateral, egalitarian relation between siblings (Mitchell 2003). I draw on Dalby's description of the attributes of geisha sisterhood to discuss interrelations within the *shisuta-seido*. This is made possible by two features commonly shared by both groups: the understanding of cultural norms, and participation in the economy as career women.[2] Sisterhood implies responsibility for the older sister who should mentor and befriend, and teach the younger sister correct manners. Shared affects of empathy, loyalty, and camaraderie preside within the relationship (pp. 5–46).

However, same-sex kinship relationships can be said to contain conflict, rather unlike the cultural ideal of opposite sex relations that contain no conflict (father-daughter, mother-son, sister-brother). Although, as Ueno Chizuko (1987a) rightly notes, harmony is unlikely given that the kin-relations based on opposite sex are cultural ideals. While there are some ideal female vertical relationships fostered by this system at JCars, there are also many exceptions where tension characterizes relationships.[3] Let us bear in mind Yuko Ogasawara's (1998: 55) brief remark that junior women are scared of senior women. As we see how the sister relation plays out between senior and junior workers in JCars, it is important that we read the intricate nuance of *hon-ne* (true feeling) and *tatemae* (formal behavior) when observing the interaction.

Mori-san is the junior "sister" to Ono-san who is in her sixth year of tenure and on the career track. Mori who started her second year at JCars is also on the career track. Mori and Ono both entered the company through *kone* and spent their childhoods in the United States. The first is not uncommon but at the outset at least, a shared experience of the socio-cultural processes of inclusion and exclusion inherent to expatriate life provides a basis for a unique bond between these women. However, rather than function to strengthen the bond between Mori and Ono, this common attribute gradually bred tension between them. They have different styles but both are talented and hard working.

I was close to Mori and Ono, and explored their sister relation through a mixture of observation and participation. By being a member of their section, I was implicated in a relationship triangle with Mori and Ono. If Mori and Ono make up the vertical axis, I was off on the side. How close or far I was to their axis depended on my role in a particular situation. As a friend, I would be close and information flow was even and circulated through all three of us. Yet, as an anthropologist, while information came from both sides, I contained it from flowing back or across to either Mori or Ono. The break in information flow showed up my awkward position as researcher, particularly as the separation of my two roles was not as distinguishable to my subjects as it was to me. Furthermore, my distance depended on the mood in the relationship between Mori and Ono. When they were on good terms the distance in their axis contracted and I was close, but when antagonized I was out at a distance as well. Furthermore, I counteracted forces that tended to pull me toward either pole in their axis. Our experience shows that our relationship was close and/or points to the peculiar dynamics of friendship in workplaces and fieldwork contexts. Now let us examine their sister relationship.

When Mori-san asks Ono-san work-related questions she often replies that she is busy and tells her to stop asking so many questions, although it is Ono's role as a "sister" to guide Mori's adjustment to her new job. In keeping with her professional image, Ono's attitude towards Mori is brash and moody, which conveys to Mori that Ono feels encumbered by the "sister" role. Mori does not acknowledge Ono's position as *senpai* (or sister) with correct greetings (*aisatsu*). Nor does Mori make it clear that she has taken on board Ono's reminders and corrections. Ono reads this to mean that Mori has excessive pride. Mori uses honorifics when speaking with her "sister" to maintain distance because she feels unable to bond with her, although at times Ono relates to her as a "friend," often disclosing information about her private life. For Mori such disclosures are unwanted because, it is not so much that Ono regards Mori as her true friend, but that Ono's position of seniority allows her to unload personal information onto Mori. In other words, like most juniors, Mori is forced to listen out of politeness. Yet Ono is making conversation to make Mori feel at ease, while at other times it is an act that attempts to compensate for how she has made Mori feel. But it also conveys a genuine liking for her. Ono experiences mild annoyance and a sense of disappointment by her "sister's" unwillingness to confide in her.

In Mori and Ono-san's group, two other women sit opposite them, one of whom, Ando-san, is on the career track. Both Ando and her colleague match Ono in length of tenure. The seating arrangement seems to divide the group into two teams; this, in fact, mirrors social relations. It can be said that the "sisters" appear to unite against Ando's team who seem particularly frosty in the mornings as Ono is often late arriving for work (by a margin of 10 minutes at most). Ono becomes very friendly toward Mori on these occasions, because Ono wants consolation and empathy. Ono often feels that Ando and colleague express general disapproval and hostility toward her, which is conveyed through a deliberate use of neutral and polite language, or alternatively in jokey banter, although their tone and facial expressions display anything but camaraderie. This might be, as Ono says, a symptom of feelings of exclusion, as they do not speak returnee-level English, or equally, it might be that their English is close to returnee standard but they maintain dignity by not openly challenging the returnee's status. I pose the alternative interpretation because, on one occasion while Ono was away, Ando said, "you could speak to me in English, too, if you prefer, since I understand" (私にも英語で話していいのよ、解るから). The tone in which this statement was rendered cast an ambiguous pall over the intended meaning. Moreover, throughout my fieldwork I encountered no evidence to verify this proclamation. Thus, regardless of their level of English, Ono is possibly correct to identify Ando's (and her colleague's) behavior with feelings of exclusion and status politics. Mori, Ono, and me sit in a row. We have many occasions to converse in English should we choose to. We might also speak in sentences that are half Japanese, half English. This linguistic freedom comes naturally to returnees, an embodiment of our history and inevitable difference from nonreturnees. Perhaps we even claim it as our right. However, we are mindful of keeping this linguistic style to a minimum as it can seem exclusionary to nonreturnees and unnecessarily disruptive, as, unfortunately, further to an inhibition or a prohibitive force implicitly felt, some nonreturnees become unsettled by feelings they identify in themselves, only misattributing it to the boastfulness (*jiman*) and conceit (*unubore*) of returnees. If Ando and her colleague wish to take part in Mori and Ono's conversation but feel censored by a lack of returnee status, or feel their linguistic aptitude goes under-recognized due to the presence and reputation of returnees, this withholding would impart residues of bitterness. The unseen barrier between individuals is mutually constructed and enforced through language. Moreover, this barrier appears to impact the level of closeness felt between Mori and Ono.

The returnee label functions in other ways. When Mori took over Ono's previous jobs, she apprehended a level of disorganization that was unfamiliar to her: it seemed that Ono was not doing the work assigned to her properly.

Thus, Mori begrudges her association with Ono on grounds of being seen to share similar characteristics through categorization as a returnee, and as her "sister." This is why Mori always speaks to Ono in *keigo* (honorifics), although Ono has asked her explicitly to drop formalities. Ono thus, without knowing why, interprets Mori's refusal to speak informally with her as a sign of arrogance. Typically the refusal to speak in *keigo* is taken as a deliberate show of rudeness and rebellion (Hendry 1990), but we can see that its opposite, the refusal not to speak in *keigo*, is equally empowering for the interlocutor as it is offensive to the recipient. Once again, we can see in this example how relationships are constructed in language use.

The tension between them also shows that Mori's status is in question: in terms of the company hierarchy she is of low status given her age and length of tenure, but her educational qualifications exceed the female norm of the two-year college diploma or four-year university bachelor's degree. Mori has a master's degree in relation to Ono's bachelor degree, and Mori's alma mater is more prestigious. Being on career track does not make her status equal to Ono's. Troubles between the two arise because Mori's experience is not recognized or validated in discourses of status. Difference among individuals is unsettling, but the tension inflates as a result of being forced to conform to hierarchical rules.

It is an unspoken rule among women in the office that junior members take the phone calls to their section. Almost everyone dislikes answering calls, particularly those from external lines. Answering the telephone has always been a female task. One male manager said this is because "a sweet female voice answering the phone is more pleasant; a man's low voice is somehow unpleasant" (かわいい女性の声で電話を取るほうが気持ちいい、男の低い声はなんとなく気持ち悪いやろ). Another male manager said he didn't like taking phone calls because people only call to ask for favors; these add to his workload.

When I began my fieldwork in the overseas department, all phone calls from external lines were answered by one section in the department that acted as the switch board for the rest of the department. Females thus answered all external calls, although each individual, man or woman, would answer internal calls through the one phone shared by two workers. Mid-way through my fieldwork the company reorganized its phone system. The external calls now came directly to each section. This system made it possible for men to take the external calls as well. More men did.

Among women it is a privilege of those in senior positions not to have to answer the phone as often. Ono-san and Mori-san, by virtue of being in a team, become the junior women in their section, but this association is forged since Ono joined JCars in the same year as Ando and her colleague. As though in retaliation, Ono always watches for others to pick up the calls. Once they take the call, a second later Ono would pick up the phone, and lament loudly in a parody of the *boke* (air-head) role in a *manzai* (stand-up comedy) act, "what a shame! I was beaten to it!" On some days Mori, being most junior, would answer over 100 calls (the daily average is 50–60 calls). Mori can see that Ono is less busy and under less pressure than her and, as Ono refuses to help, Mori knows that Ono is upholding the vertical relationship. Both sides related other incidents to me that caused tension between these women in the tutelage system, but further details are unimportant to my argument. To conclude, Mori requested a transfer after her first year, to another section of the department (which has its own room), primarily to further her career interests, but also to make a fresh start among new members.

Once, Ashikawa-*jichō*, our section head, interceded voluntarily in order to restore fairness among the ranks. While Mori was away from her desk Ashikawa asked the senior women, Ono and Ando, to "please, won't you take the calls [for Mori-san]" (電話とってあげてね). Ashikawa is of kind and gentle disposition, which, combined with his use of diplomatic and placating language, introduces the feeling that the seniors are doing the juniors a favor. This intervention, reinforced by the effect of not directly naming the subject, allows Mori to save face. Nonetheless she suspected the manager's hand because there was a noticeable change to the senior's behavior. Relations between women become complicated by hierarchy, combined with the frustration of feeling undervalued by the gendered task they are required to perform. Mori and Ono's volatile relationship is quite typical; I observed similar tensions within the sister relation and outside it in women's relations with other women. I note, however, that women and men are reluctant to speak about vertical relationships (これは栗原さんにはあまり言いたくないなぁ). Such reactions testify to the difficulties within them.

My experience of working on the English language company newsletter (Company World: CW) circulated globally to clients, affiliated companies, and subsidiaries further illustrates the way the company and sister hierarchy operates. Ashikawa-*jichō*, our section head of the overseas department, was in charge of selecting articles that appear in CW. Articles that first appear in the Japanese language issue of the internal company newsletter (Company News: CN) are translated into English. These are then included in editions of CW. The articles are selected based on their news-worthiness to overseas subsidiaries and to external audiences, particularly if they are up-to-date. However, the inclusion of any articles relating to company performance is decided by a higher ranking board member. Four editions of CW are published annually.

Returnees (bilingual in Japanese and English), the junior workers, translate the articles. I am included in the tasks as well. We spend long hours working on the difficult translations, often leaving aside other jobs. The articles are heavy on technical jargon. Once translated, the work is faxed or emailed to Burns-*kachō*, an American section head, at a subsidiary in the neighboring Prefecture. Burns' job is to edit out all or most of our writing so the piece is unrecognizable. We feel as though our linguistic competence and writing skills are considered poor and redundant. We interpret the negation of our work as evidence of the function and consequence of traditional hierarchical structures operating within the company. We discuss among ourselves that the company gives us translation tasks in order to generate a feeling of usefulness among junior workers. For this reason these tasks are not received neutrally.

The editorial side of this operation is taken on by Ono-san, in which she has three years' experience. It is her role to plan the layout of the articles and photographs. Often locating the photographs is cumbersome, as they are stored among various departments and at overseas subsidiaries. This is a sign of both disorganization and of distributed organization. She schedules a meeting with the printing firm to go through the specific requirements for each upcoming edition. The photographs and the disk on which the articles are saved are handed over ceremoniously. A week later the publisher sends the test print to Ono. After further editing by Burns, the second draft is received and on the strength of this the official version is commissioned. Ashikawa scans the second edition, as Burns has exercised thoroughness beforehand. Ono facilitates this work process, and her role is clearly important. The whole production cycle from the selection of articles, translation, editing, to completion, takes roughly one month. The process involved in the preparation and publication of CW, as described above, highlights clearly the formulaic way in which young workers are given opportunities, only as practice or training for future roles if, or when, they reach senior positions.

Ashikawa suggested that I share the responsibility of editing CW with Ono. This was Mori's job, but, as she transferred to another section, it came to me. Ono then became my "sister." It was by becoming involved in CW at this level that I was able to learn more about sisterhood and understand Mori and Ono's relationship better. Furthermore, I was able to empathize with the feeling of junior workers whose decisions and suggestions are overridden by managers higher up in the hierarchy (whose job it is to maintain tradition).

Like other junior workers, I experienced some uncertainty when I first began this task. I had enthusiasm and some innovative ideas to transform CW into a more readable, enjoyable, and useful publication by changing the tone of the articles and incorporate different colors and fonts.

Taking the appropriate channel of communication, I first speak to Ono as she is my "sister." "No," her immediate response: there are rules, it's been decided, it's always been this way, and it has to stay the same.

Soon after, when Ashikawa spoke to me about some excellent changes suggested by Ono, I realized, the extent to which hierarchical norms operated in this relationship. Clearly, by recourse to traditional notions of the workplace the senior worker was able to exaggerate to the junior worker the difficulty involved in making changes. With her longer tenure and greater experience, Ono knew exactly which changes to propose to the manager.

Perhaps change is often instigated by external sources. Taking my ideas on board, now Ono wanted to make those changes. I had been naive to the fact that competition is an inherent part of our relationship (our similar background and status as returnees; our common linguistic skills; our fathers' positions within the company hierarchy—although differing by age, they occupy similar posts, received early promotion throughout their careers, and significantly increased profit margins). Because Ono is my senior, I thought we would be exempt from a relation of competition. This is when the fact that "sister" ideals are not reality was impressed upon me most strongly. I approached Ashikawa from thereon with my ideas. But he is satisfied by the standard of the CW as it is. It is not within his or JCar's interests to make CW more interesting or to change it radically.

After assisting Ono on one issue of CW, Ashikawa suggested I try out the editor role. I would be able to include a wider range of articles if I could find interesting ones in industry magazines and economic publications. When this handover of tasks was appointed, Ono sent round a departmental email to say that I would assist her officially, rather than that she would assist me unofficially. By this point I judged the need to assert myself; thus I noted this factual discrepancy in a voice that all members of the section could hear. Ono amended her statement in the next email that she sent to Burns. This interchange of roles gave rise to further tension in our hierarchical relationship (as she also defined me as a friend). I requested her input but Ono made it obvious that she did not want to be involved. As I was in charge, she said I should know the requirements of the job. I thought it unjust of Ono to react in this way; as it was by her own request that she gave up the editor's role.

Eventually, Ashikawa comes round to "our" idea of changing CW after having a look at a competitor firm's newsletter. He walks over to our seats and shares his thoughts: "theirs is much more professional looking than ours." He is embarrassed to show ours at the important, annual general meeting of the overseas heads of subsidiaries that he is responsible for hosting in the coming month (December). I feel sympathy for his position. He agrees that CW could take more stylistic risks in order to ignite the readers' enthusiasm and interest. We don't have sufficient time to make a new edition for the general meeting. And my fieldwork will finish before the next issue comes out. At times hierarchical social relationships and traditional work routines obstruct innovation, and it envelops everyone at all levels of hierarchy. Yet, from the co-ownership of new ideas and realignment of objectives that emerged organically, we can also observe a sense of community and purpose being reformed within the group.

We should be encouraged by such outcomes but this is not always the case. Miura-san has been working for one year doing clerical work in the accounts department. The inefficiencies of traditional work methods are clear to him. However, as tradition does not allow for change, he is unable to alter the aspects of his work routine to enhance his productivity. To do so would interfere with other people's work as many jobs are linked within an interdepartmental process. The colleague in another department who passes on his work to Miura makes many mistakes and causes delay and confusion in his job. If he has access to the original files stored in my section, he could by-pass this problem. I could easily include him in the list of people I send the files to, but organising this change requires permission from his boss. This bureaucratic hurdle puts Miura off. He experiences similar dilemmas to mine in his relationship with his direct manager. Miura's attitude and enthusiasm toward his job suffers as a result. He is pragmatic but never cynical.

Kasuga-san has been working for one year in the overseas department in a different section to me. Both she and her "sister" are on the career track. One day her sister says, "I bet you have nothing better to do, so why don't you write this letter that needs to be faxed urgently. I was asked to do it but I can't be bothered. It's important: the addressee is someone very senior in another company, so try not to get it wrong." The task takes the whole morning to complete. When her sister looks through it, she criticizes Kasuga's style for being long-winded and full of unnecessary phrases. After some argument between them, the sister deletes the two lines over which Kasuga had labored the most. Kasuga thinks the two lines are crucial because they carry the sentiment of the letter; the feeling is important in this case rather than the details. The sister hands Kasuga's work in to their boss, claiming credit for the work as her own. The letter is returned to the sister with a handwritten note, which reads "Inaccurate. Redo. Include [the part that the 'sister' deleted]." The sister laughs (to mask her error in judgment and to maintain her higher status) and tells Kasuga to rewrite the letter. The sister had erased the original letter thinking that the submitted version would be accepted. It is 5PM and Kasuga is feeling used and annoyed. She considers leaving a note for her sister asking *her* to redo it and go home. This is what she would like to do but does not dare, so she stays to finish the work. This fax must be sent before the end of the day.

Unlike the relationships above, Hori and Igarashi-san have the ideal sister relationship. Although Igarashi is engaged to be married and has a lot on her mind, she makes sure that Hori is not struggling with any tasks, and makes time to clarify questions. This kind of relationship is rare. Is it a coincidence that they are sisters in the personnel department whose members are expected to act as models for the rest of the employees? And that Hori was one among many who did not want me to ask about her experience of vertical relationships? In this respect, it is not surprising that their relationship embodies the ideals of the company.

The examples above suggest that hierarchy is a unifying principle at a conceptual and symbolic level, that is, at the level of the official discourse of the company. Moreover, hierarchy itself does not engender the qualities of harmony. There is no doubt that structure is one of the fundamental principles of social and cultural constitution. Hierarchy functions to express the unity of the cultural system through "reference to its social values" (Dumont 1980). But these social values are descriptions of an ideal. Although the management envisages hierarchical relations between "sisters" to function as a nurturing type of relationship just as Nakane (1970) explains, my experience during fieldwork and observation of other workers' experiences relates an altogether different experience. We were able to account for this by looking at the way relations are constructed in language.

Work is hampered by hierarchical tensions among its members and workers use hierarchy inadvertently and deliberately, at different times. Nakane (1970: 14) also notes that the authority of individuals in higher positions of hierarchy can abuse the group or members, but this is allowed because the relationships are fundamentally characterized by dependency and loyalty. In Mori and Kasuga-san's experience and also through my own, the individual in higher positions within the hierarchical relationship assumes that juniors should have feelings of loyalty and affection, and this coincides with the principle outlined in Nakane's model. But these are feelings that do not automatically emerge via the membership to a group or by assignment to vertical social relations. These underlying assumptions regarding the nature of hierarchical relationships are contested and questioned particularly by the workers in junior positions. This goes to show the degree to which the nurturing sister ideal at JCars is one-dimensional, abstracted from, and is at odds with reality.

From a psychological perspective Terri Apter (2007) shows that sisterhood, like friendship, is characterized by a profound ambivalence toward the other; existing in relation to the positive feelings are resentment, envy, hostility, competition for status, and anxiety around losing our place in the world. When work colleagues are placed in relations that carry a familial metaphor the same dynamics can play out.[4]

In addition, relations between co-workers are strained because each side, both senior and junior, know that these relationships are ideals. And, therefore, this knowledge of ideals is in need of reinforcement to make it real, which is why monitoring of affect is important within these various hierarchical relations at all levels within the company. As we can see, within hierarchical relations junior workers are made to behave in ways that guarantee and display the senior status of the other. If juniors don't show their deference consistently, the relationship can malfunction, or there is a channel of dissatisfaction coursing through the relationship. In the context of relationships with other *senpai* outside of the sister relationship, the hierarchical rules of behavior can become enforced more strictly, thus daily life can become more challenging for junior workers.

For example, if a female *senpai* is making her way round the department with a box of treats (often cookies or chocolate) in late afternoon, even if the new female recruit (bottom of the hierarchy) is offered some, the junior member must override their real feelings and refuse. The junior must read the *ura* (back, in relation to this showing of *omote* [front], because the two exist simultaneously) and say "thank you, but you must have it, please, *senpai*" (結構です、先輩どうぞ). This is how hierarchy is displayed and reinforced actively in language. This symbolic exchange, too, forms part of a collective practice of maintaining solidarity within social relations, what Jacques Lacan calls " 'empty gesture,' an offer made or meant to be rejected," with distinct social gains—if none material—for both parties (Žižek 2006: 12–15). The practice of grasping the distinction between *ura* and *omote* is universal.

The junior woman must also be aware of the *senpai*'s or groups' needs, otherwise she loses face. For example, women of Ono-san's section in the overseas department are called on to serve tea at the important annual meeting where heads of subsidiaries report back to the head office (as mentioned above). The meeting is scheduled to start at 1PM. The *senpai* who is coordinating the preparation requested that everyone gather at the tea room at a quarter to 1. Ono is on time but all her *senpai* were already there. The preparation was at its end stage. The *senpai* responded politely but coldly, just enough to display their disapproval. Until that experience, it did not occur to Ono that she should have gone earlier. Afterward she emailed her *senpai* to apologize for not being more aware of the group's needs. She received a cool reply saying that others helped so it went fine without Ono's help.

Let us now consider the interrelationships within *dōki* groups.

Dōki Groups

[A] horizontal tie would be established between X and Y, who are of the same quality. When individuals having a certain attribute in common form a group the horizontal relationship functions by reason of this common quality. (Nakane 1970: 24–5)

「持つべきものはいい同期」、という企業社会の "伝統" が過去のものになりつつある。 というのも若手社員の間では同期の仲間意識が薄らいでいるのだ。 終

身雇用が崩れ、能力主義の導入や中途採用が相次ぐ中、人間関係もかつてとは変わっている。

["One must have a good *dōki*," this "tradition" of the enterprise society is becoming a thing of the past. That is, between younger workers the sense of friendship between *dōki* members is waning. With the breakdown of the lifetime employment system, amidst successive waves of implementation of a merit based promotional system and mid-term recruitment, human relationships are also different from what they were in the past.] (*Nikkei* 02/02/98)

In this section I examine the reality of relations within *dōki* groups compared to the ideals that structure them. I examine the contrast between the discourse of management and the workers' experiences of *dōki* groups. When a worker joins JCars they are told that the *dōki* group is an important source of moral support. This is because each individual is seen to be going through the same learning experience as the others in the group.[5] The group is united by the common year of entry into the company that is fixed by the management; this falls on April, as this month marks the official entry into schools as well as companies. However, this unifying factor does not apply to all recruits. *Chūto nyūsha* (mid-term recruitment) describes a recruit who enters the company at any stage past April and *tenkin* describes those who moved to the head office after having been initially recruited by another subsidiary in Japan (this relocation rarely applies to women at JCars). Both *chūto nyūsha* recruits and *tenkin* workers, for whom there is no set date for entry, become allocated to the *dōki* group at the head office whose members joined in the preceding April. Often, *chūto nyūsha* recruits are slow to share the same degree of bonding with their *dōki* group as they would not have attended the new recruits induction seminar as the others have done.

The induction seminar for new recruits in April is followed up by another two-day seminar in July. Oda-*shunin* from the personnel department, who is in his thirties, is the coordinator. This seminar aims to "build friendships" (仲間づくり), "foster the unity/harmony of people" (人の和を育てる), and to "construct human relationships" (人間関係の構成). Oda said these aims are difficult to realize, as the recruits who join the company by *chūto nyūsha* participate in the July seminar together with the majority of recruits whose group already bonded in April's week-long seminar. This makes the running of the seminar difficult for him. Clearly, the personnel department in this accord are attempting to create and reinforce idealistic notions of the *dōki* group despite recognizing some inherent difficulties.

Dōki groups are a source of information exchange, so the function of the group is useful for gaining wider knowledge about the company, where insight into the organization of different departments and managers can be learnt. Baba-*shunin*, a mid-ranking manager, said that members of his *dōki* group are "trusted friends/fellow colleagues in whose presence one can drop one's guard" (気を許せる仲間). As all of the women in his *dōki* group have retired, he is referring strictly to relations between men.

But there are elements of competition that arise among the group the longer they stay in the same company, as some *dōki* members are bound to be promoted before others. As long as the *dōki* members in competition are not in the same department, the friction arising from the subsequent status difference between *dōki* members may not matter to the functioning of the overall group, for the vertical relationships between seniors and juniors, and not the *dōki* group, matter more under competition for promotion. Men's promotion very much depends on their boss's power and influence. Baba's views of the *dōki* group is closely based on the ideals of the management discourse, resembling Nakane's views discussed above. This is also echoed in the account below.

Kuwahara-san's view is representative of the younger male workers' views on *dōki* groups in the head office. Kuwahara-san is in his fourth year at the company. He had been transferred to the head office from the Tokyo office one year previously. He has very few friends at work, as there are only three people from his *dōki* group (he only accounts for the men in this group) at the head office. To Kuwahara, *dōki* members are people with whom he has spent the greatest amount of time since starting his career: for six months prior to being allocated to a position in their respective offices, male recruits go through an intense period of training in one of the company's manufacturing plants (*kōjyō-jisshū*). The junior men socialize in the *dōki* group after work (eating and drinking) and at weekends, mainly because most of them live in the same company dormitory and are unmarried.

However, this is not the only reason why men socialize together more frequently in their *dōki* groups than women (discussed below). It is important for men to nurture this social bond because these men are likely to work together in the same company for the rest of their careers, although interfirm mobility is increasingly becoming socially acceptable and an economic necessity. Generally, and comparable to men, women, who are likely to retire from positions in large companies following marriage (although currently many women push back their retirement), express a less straightforward and uniform opinion of *dōki* groups. On a similar footing to the accounts of men outlined above, the women's views of their *dōki* group mainly refer to social relations with other women. Although men and women of the same *dōki* group socialize together, opportunities occur rarely, therefore *dōki* interactions refer more to same-sex members. So how do women experience *dōki* interactions?

Shimoda-san whose career is in her ninth year chose to apply to JCars because the intake of recruits was large (i.e., a large *dōki* group): "the number advertised for recruitment was close to 30 people, this was a lot, so I thought it might be more fun." (募集人数が三十人ちかくもあって、多かったから そのほうが楽しいかなと思った). Takatomi-san said that with a large *dōki* group smaller cliques develop over time that split off once individuals form bonds through shared interests and personality type. This results in those members of different smaller cliques or factions (*habatsu*) exchanging only formal greetings. Thus, Takatomi is pleased to be in a small *dōki* group of five women because it is easier to be with like-minded people. Takatomi sees the women in her section as colleagues who are "associates by work only" (仕事だけでの関係). Invariably, as women in her section changed over her three years of tenure, Takatomi considered her *dōki* relationships more important than female relationships in her section as there is stability to the membership.

Nakai-san who has been at the company for five years said: "With regard to friends, I am closer to the friends from work and often go out with members of my *dōki* group. I only exchange New Years greeting cards with my friends from junior college" (友達に関しては会社の友達との方が親しい、よく同期の人と遊びにいくし。短大の時の友達には年賀状を出すくらいの関係). Thus, Nakai's friendship with people from the company is more important and immediate than her classmates from junior college. Makimura-san, who has been at the company for two years, said she valued the relations with her *dōki* members above all other social relationships in the company. Often, *dōki* members share the same feelings about the difficulties of the working environment and particularly, the members of her group are willing to listen to her complaints. To cushion the nervousness she experiences when she makes errands to other departments, Makimura approaches her *dōki* member who then makes an introduction. The *dōki* group is synonymous with the company for Makimura who said, "I tend to view my *dōki* group as the company" (同期を会社という目で見てしまう). Although Makimura said she "feels greater intimacy [in her relationships with the *dōki* group members] compared to [the previous year] when she entered the company, she does not feel compelled to know everything" about a particular person in her group (入社のころより親近感は増したけど すべてを知りたいとか思わない). Clearly, the *dōki* group is important to Makimura primarily in terms of emotional support, but the maintenance of emotional distance suggests that Makimura conceptualizes working life and private life as two quite distinct spheres.

Referring to the vertical relations existing within the *dōki* group, Nishida-san said that age differences indicating women's varied educational backgrounds (a junior college graduate is two years younger than a university graduate) does not affect language use (i.e., she does not use the polite form of address toward older university graduates if they are in the same *dōki* group). She goes out with some of the members of her *dōki* group after work where she exchanges gossip and information about shopping and men. Nishida said: "The girls in my *dōki* group are like my friends from my student days" (同期の子は学生の時の友達みたい).

Mori-san, who we read about previously, had ambitions to study abroad for a doctorate or MBA. Instead, she followed her parents' wishes and entered a traditional career path at JCars. For Mori, the most difficult aspect of working life is negotiating successful social relationships, particularly between women. From the first, people are forced into "a group created at will by the company" (会社が勝手に造ったグループ). Mori said that such a group created purely on ideals obviously functions to create a sense of unity or harmony (和, *wa*) precisely because such "unity is in disarray" (和は乱れている, *wa wa midare te iru*) or did not exist among Mori's *dōki*. For Mori, the *dōki* group consist of "social relations that are maintained and endured out of obligation" (義理の付き合い). The group has little significance to her, but she conforms to the rituals of eating lunch in the *dōki* group to satisfy company conventions. On the few occasions when Mori had broken the unspoken group rule and went to a café on her own she faced the reproach from members of her *dōki* group. They commented that Mori is "strange" (*kawatte-iru*). Mori then decided it is better to feign the appearance of liking the company of her female *dōki* group.

In my final week in the field I came across a surprising observation. Mori took a call from her mother on her mobile on our way to dinner. Mori spoke in Osaka-*ben* (dialect)! This is noteworthy because Mori only spoke Tokyo-*ben* at work. Until that moment, neither I, nor anyone else in the *dōki* group, section or JCars, had heard Mori speak in Osaka-*ben*. She doesn't like Osaka-*ben*, and prefers Tokyo-*ben* since she feels it expresses who she is. Her *dōki* group makes fun of her Tokyo accent, but presumably this is because it symbolizes her life experience and superior educational attainment that serves to differentiate her from the *dōki* group. Dialect carries meanings of status difference between rival cities, Tokyo and Osaka, as well as the metropolis and rural hinterland dichotomy. Contrary to popular belief it is not that Mori couldn't or hadn't mastered the use of Osaka-*ben*. In fact, she was born in Osaka and lived there until relocating to the United States at the age of 11. Similarly, in a friend of mine's home Osaka-*ben* is banned from use because her mother, whose provenance is Osaka, prefers the refined air of Tokyo-*ben* to the coarseness of the former. Her father and sister get a telling off when overheard on the telephone speaking to friends in Osaka-*ben*. This shifting of dialect use between inside and outside of the home, that is inconsistent with actual geographical location or institutional context, is rare (according to my subjects), but not unusual. Language as symbol of individuation and emotional distance from others comes alive in Mori's example.

Araki-san, like Mori, expresses a similar dissatisfaction with the "unnatural" *dōki* group concept and disagrees with the management's tactic to generate camaraderie. Araki links her experience of *dōki* group participation with issues of gender segregation:

会社でなんでも同期にしたがるコンセプトが嫌い 。。。 仲良くしないといけない
雰囲気 。。。 他に気が合う子がいるのにごはんを同期と食べないといけない。
同期なんて新入生導入研修で仲間意識をつくり 仕事に有利にしようとしたコンセ
プトとしか思えない。 男の人は別で総合職の研修があるから寮に入って 二人部
屋に三ヶ月間住んで 自然と仲良くなる。 そうもすれば 同期で知っている人だし
仕事を頼まれたりしたら まー やったろうとも思える。 女子は総合職の研修は無
い。 女子の関係も同期とするなら 男子と別に扱わないでほしい。

[At work I hate the concept where anything is eagerly oriented around the *dōki* group...the atmosphere of having to be friendly...I still have to eat lunch with my *dōki* although I get on better with others in the company. The *dōki* group is nothing more than a concept created at the time of the induction seminar for new recruits that generates a feeling of camaraderie which can be used advantageously for a work-related purpose. It is different for men who become friends naturally, through attending a career track training seminar and by living two to a room for those three months. That way of course one knows their *dōki* and if they ask you to help them in a work situation you would think; yeah, sure, I'll help ya. Women don't have a career track training seminar. If the management is going to apply the *dōki* concept to women, I wish women weren't given separate treatment to men.] (Araki)

Shimada-san has been at the company for 10 years and most of her *dōki* group has retired. She does not consider the two remaining women from her *dōki* group as particularly good friends. Shimada eats lunch with Shimoda-san, a woman from her section whose *dōki* group is small, as Shimoda prefers not to have lunch with them. Shimoda is Shimada's junior by one year; they work in the same section, and, through a shared interest in tennis, they have become friends in and out of the office. Most of Shimoda's *dōki* has left the company as well.

Matsudaira-san feels that developing close friendships among *dōki* members narrows one's network of social relations: "if my good friend at work should leave, I would feel uncomfortable staying here" (会社に仲の良い人がいれば その人が辞めたとき 自分が居にくくなりそう).

When an employee retires, farewell parties are organized by fellow workers of the individual's section, department, and the *dōki* group. Li-san, who by nationality is Chinese, and lately feels neither Japanese nor Chinese, studied at a university in Beijing. Two years into working as a travel agent, tourist numbers fell following the Tiananmen Square tragedy. Li's father is a science professor at the prestigious Kyoto University, and while staying with him she studied at a language school. After 1½ years, Li passed the university entrance standard examinations and decided to find work in Japan. Her experience of these parties organized by female *dōki* members is that "very few women participate in leaving parties with feelings of sincerity" (心から思って送別会に参加している女の人は少ない). Thus, Li feels that in the majority of cases these women attend out of feelings of obligation toward the leaver. The fact that on the first day Li arrived to work in the office she was invited to a woman's leaving party made her question the nature and genuineness of these parties. Li said all that happens is people eat together and take photographs. This attitude "does not constitute a meaningful exchange of communication" (交流になってない) between participants that caused some consternation to her, but she said the others seem happy with this.

While the *dōki* group is a manufactured concept based on harmony and ideally it functions smoothly, the *dōki* group itself can become the source of dissension as the following example shows. Igarashi-san recalls how her relation with Ota-san became strained when Igarashi's male friend (another colleague) to whom Ota was attracted got married. According to Igarashi, Ota developed a grudge against her and responded by forming a clique with others in the group to ostracize Igarashi. Ota would ask everyone in the group except Igarashi to go out at the weekend, and this situation continued for a few weeks. The other members of Igarashi's *dōki* would not openly criticize Ota's behavior, although they gave Igarashi emotional support in private.

I observed this *dōki* group when I ate lunch at my desk. The group often ate in silence. Their politeness implied *tatemae* behavior. Once I was invited by Igarashi to join their group for lunch so that I could conduct an informal group interview. On this occasion too, the members were equally subdued; although I had wondered if my presence had altered the group dynamic; this was not the case. Igarashi said they ate together despite having nothing in common, in fear that one member of the group would be left alone if the members chose to have lunch with people or in groups of their choice. This moral consideration functioned to maintain the structure and longevity of the group.

Summary: Ideals and Experience, Uniting Theory and Practice

The various comments quoted above relating to the sister tutelage system and *dōki* groups indexes the idealism supporting the nature and function of social relations within a company founded on group-based ideals of harmony. These ideals are not synonymous with qualitative characteristics occurring naturally within social relationships in the office. The multiplicity of interpretations conveyed above contrast with the static and straightforward representation of vertical relations and *dōki* groups often proliferated by both management and in models of Japanese social organization. The grouping of workers by year of entry and, moreover, the essence of harmony upon which the concept is founded is experienced as unnatural and banal in most cases, which exemplifies the extent to which harmony in the workplace is an ideal that is uncharacteristic of reality.

The *dōki* group is part of the structuring principle used to socialize the recruits into a new environment of complex social relations. Ironically, it seems to complicate them further. The functionality of the group seems more important for men (cf., Matsunaga 1995; Rohlen 1974), who conceptualize and experience the group straightforwardly in ideal ways, and appear to accept it unquestioningly unlike the women who express greater variance of opinion. Perhaps men are more pragmatic and better able to assimilate the rules. Or rather, female workers are more concerned to question and distinguish between different degrees of friendship than male workers. Women may have a stringent delineation between work and friends at home compared to men of the same age. That said, for some men and women, and this boils down to personal perception, the *dōki* group is useful in the work-related context of interaction whereby workers in the group are genuinely regarded as friends.

The examples above, which highlight the dissonance characterizing some of the relationships between women in hierarchical relations, contest the main thesis of structural-functionalist analyses of the workplace based on harmony and consensus that again is

> The vertical relation which we predicted in theory from the ideals of social group formation in Japan becomes the actuating principle in creating cohesion among group members. (Nakane 1970: 26)

There is little cohesion experienced by workers in both their vertical and horizontal relations. But cohesion exists as an ideal in the minds of workers by virtue of being placed within these social structures. And the appropriate behavior is called upon to reinforce these ideals, to make them real. This is true of both working relationships and lunch groups. As seen in the ritual forms of lunch groups, the vertical principle of social organization likewise requires a performance in an enactment of ideal forms of behavior to demonstrate the knowledge of social structure. This enactment of form creates a sense of cohesion and harmony:

> A strict hierarchy between OL's is shown by the deference *kōhai* paid *senpai*... Of course, other factors besides tenure affected the deference shown to a person. For example, how well one knew a person was also taken into account: the more intimate one was with a person, the less elaborate the formality. (Ogasawara 1998: 48–9)

Echoing the quotation above, at JCars, the role of the junior was to serve (*tsukusu*) the senior and the role of the senior was to teach (*oshieru*) the junior.

Sentiments and behavior expressed in ways as shown in the example below maintain the characteristic of harmony within interpersonal relations in the workplace. Hori-san limits her conversations to senior and junior workers to work-related topics. If the conversation concerns general topics, Hori (as the junior individual) restricts the conversation to formal topics, taking care to avoid personal topics that might make the senior individual uncomfortable. Equally, Hori does not mention inconsequential or inappropriate topics in such situations, as she does not want rumors about her private life to be spread about the office. Discussing neutral topics that have no recourse to the relationship is important to Hori from a perspective of reciprocity, as a revelation contains the expectation of an equally valid revelation. The negotiation of *hon-ne* and *tatemae* by the individual is the key issue in the negotiation of vertical relationships. This also extends to horizontal relationships as well. Hori's account is a general rule of conduct within workplace relations. The extent of the care taken to select topics of conversation marks the degree to which an individual works to conscientiously manufacture and maintain idealistic social relations.

Thomas Rohlen (1974: 106) likewise casts doubts about the ability of hierarchy itself to maintain harmony and solidarity within the group. He finds that the group cannot be described in terms of having only one structure and if one looks at different contexts of interaction, such as work and office parties, the organization of the group changes from a pyramidal to a circular structure, and it is the role of the group leader to instigate this. In Rohlen's view, harmony at the level of the company is an ideal, but it is a reality at the level of smaller groups (sections). Relationships in the section are essentially harmonious although they can sometimes stray into discord and dissatisfaction, and it is in these moments that its members experience the hierarchical structure most strongly.

Nakane (1970) and Vogel (1963) describe white-collar office workers' relationships with each other as true friendships. But this is an overstatement as Reiko Atsumi (1980) points out: a worker's relationship at work is often different to friendships, more often they fall under *tsukiai* (range of personal relationships) that indicate feelings of obligation and social necessity. So it is clear that Nakane and her critics are pursuing very different lines of enquiry. Whereas Nakane is interested to know the general properties of hierarchy in abstract terms, Atsumi and Rohlen are looking for facts grounded in everyday life (also see Goodman and Refsing eds. 1992). They show us that by accepting the abstract we miss the real issues. But we also need the abstract as guide. The group model, then, like all models, is best described as a "folk model" rather than as a rigorous "scientific model" (Befu 1980).

Put differently, hierarchy is imbued with the difficulty of describing behavior from a dual perspective, that is, the theory of rational individual choice based on self-regarding motives, and a notion of collective behavior witnessed in organizations (Douglas 1986). Mary Douglas's point is that the holistic structural-functionalist explanations of social organization sit uncomfortably with the simultaneous consideration of individual action and collective behavior. Let us then reposition Douglas's point in relation to Japan. Researchers are criticized for their inability to distinguish between *tatemae* and *hon-ne*: that is, between "values and voluntary behaviour" and "ideologically manipulated or coerced behaviour" (Mouer and Sugimoto 1980). Indeed, distinguishing between the two is difficult for researchers because subjects work hard to display *tatemae*; they do this because ideals are valued and/or to protect the positions from which they speak.

Distinguishing between *tatemae* and *hon-ne* is complicated because, as we know, it goes hand in hand with understanding their meaning. Cultural meaning emerges from understanding how people use ordinary language. And meanings emerge in contradistinction to language use. This is noticeable in situations calling for enactment of ideals. As we have seen, *keigo* masks contrary feelings and functions as a social barrier; responses don't express real feelings but instead align with hierarchy, maintaining an artifice of solidarity and rank through empty gestures; dialect individuates, liberates, and empowers, while expressing feelings of estrangement from the group without recourse to words that formalize the underlying sentiment. Feelings, like meaning, are unstable and ambivalent, colored by different moods. In these subtle and supremely incontestable tactics of distantiation we observe the mystique of language in its innumerable guises, and there is more.

Silence, for one, is artful communication: empathetic, sensitive, and intuitive (Kondo 1985: 304; Lebra 1976: 116, cited in Kondo 1990a: 325, n. 12)—silence underscores a sense of togetherness and consensus in Japan. A silence, meanwhile, holds together an illusion of oneness; for example, in the context of the bedroom, a pregnancy ensues from a hesitation to suggest contraceptive use (Riley 2005: 71–83). Following Denise Riley's analysis of words, affects, and meaning, can we not say that *both* types of silence are behavioral-linguistic adaptations that circumvent the capacity for linguistic violence and its unknowable effect on social relationships? Words, then, are necessary to establish separateness but risk disrupting ideals. At times subjects cannot "talk" in silences; vocal responses are often mandatory in the workplace. For instance, we observe conversations where subject's responses display little correlation to what is said: the subjects talk out of time, over each other's heads, and fail to complete the circle of reference. Despite a mutually felt empathy, workers must nonetheless express divergent opinions that provoke discord. Hence the indirectness.

Language complicates relationships. Many of my subjects, irrespective of gender, directly connect the complexity of their own workplace relations to such difficulties of the Japanese language. Their remarks are free of culturally specific *nihonjinron* (theories of Japanese uniqueness) overtones. If Japanese nationals find their language difficult, would not researchers also? Presumably, but obviously, this statement has negative implications for a researcher. This taboo is worth lifting. Looking at what we understand less well will give us access to new knowledge.

In organizations there is a lot of *tatemae* talk; it is often the way people speak to each other (including the researcher), always carefully delayed, veiled in politeness, indirect, cryptic, vague, multilayered thus open to multiple interpretation. Meaning dissolves and steals away at the moment you think you have grasped it. In large part this stems from the structure of language itself, which does not provide many clues as to how the conversation is developing: in addition to the problem of the omitted subject, verbs appear at end of sentences leaving room for change of direction depending on the reaction of the other (Pascale and Athos 1981, cited in Nonaka and Takeuchi 1995: 53, n. 29).

Ways of speaking (*ii-mawashi*) create complex meanings that range from difficult to the impossible. As mentioned above, periphrasis (*tō-mawashi*) places distance between the utterance and the subject, enabling both parties to save face; for a fleeting moment meaning lives in time, like smoke, time stands and swirls, but in a moment drifts away. If our subject's mutual incomprehension alienates them from the present moment, putting them in a fictive relationship with each other, through language, they perhaps imagine the other, as, does Roland Barthes, a "great envelope empty of speech" (1982: 7). Time, then, surrounds the empty subjects, allowing ambiguities and contradictions to settle, a resolution of unresolved meaning. This shows the opacity of language as possibility and its limit. It is as though we tread softly within our social relations, reading their meaning like a haiku: "to suspend language, not to provoke it" (p. 72).

Language on its own can be more difficult to understand than behavior, but distinction between speech and behavior is unnecessary. Following Judith Butler (2004), we can think of language as presenting the body in its action. In other words, saying is a mode of behavior. Working in a company (confinement of the body to a chair and desk, with strict rules of hierarchy in operation) is not at all like being on the analyst's couch, but if they have anything in common it is that the subject articulates/acts principally through speech, for the analytic situation:

> . . . does not put the body out of play, but it does enforce a certain passivity of the body, and exposure and a receptivity, that implies that whatever act the body will be able to sustain in that position will be through speech itself. (Butler 2004: 172–3)

This applies equally to the interview situation where the researcher can, for the main part, only work from the subject's positions and feelings that are enacted through words. Language use is important, as Denise Riley (2005: 3) writes "How Words Do Things *with Us* [my italics]...as distinct from 'to us' " is paramount:

> Laments, rhetorical questions, exonerations, comedies of verbal inhibition, and clichés...these all exert themselves as ordinary effects which are, though, no mere embellishments or overtones on top of their speaker's intentions: they can even outrun them. Or they can make their speaker's sentiments virtually irrelevant. (Riley 2005: 4)

As we saw from the example of the contraceptive glitch above, Riley aims to get beyond ideology by looking at uses of ordinary language and the affects that work through it (p. 6), for instance, the disruption caused when semantic meaning is pulled away from the nature of the utterance and from fact (p. 75). Meaning is not straightforward; this is why understanding it is interesting.

As we covered in this chapter, observing the idealistic formation and function of *dōki* groups and vertical relations at JCars give us reason to approach the Japanese organization not from a perspective of harmony and consensus, but from assuming dissensus as the main characteristic of social relations from which the organization or management must form the ideological and cultural construction of the workplace. Trust and distrust (see Schindler and Cher 1993), "parasitic," "accidental," "one-way," "mutualistic," and "transcendental" (Hicks and Gullet 1975) are alternative ways of experiencing social relations in organizational contexts. We have seen that harmony is not an intrinsic character of the workplace, nor endemic to vertical and horizontal relations. We also saw the way at times hierarchical social relationships and traditional work routines obstruct innovation, and it envelops everyone at all levels of hierarchy. Yet, from the co-ownership of new ideas and realignment of objectives that emerged organically with time, we observed a sense of community being reformed within the group. In the next chapter, we will see how the management discourse constructs an image of community and enforces ideals through symbolic means.

CHAPTER 6

Temporal Dimensions of Symbolic Community

The preceding chapter examined the form and function of the sister tutelage system and other supervisory relations as well as interrelationships occurring within groupings of workers organized by year of entry. In the picture that emerged, we witnessed the complex landscape of emotions and linguistic practices contrary to the discourse of harmony so vital to organizations. In other words, we located the discrepancy between the subject's experience of vertical and horizontal relationships and the ideals by which they should function. Equally, we saw how some workers perceived relations to be based on genuine friendship, which was also useful in helping each other in a work-related context. The question that concerns us is this: if harmony is not an intrinsic characteristic of the workplace, nor endemic to vertical and horizontal relations, then what countermeasures operate to ensure the continuance of a community resembling its ideal form? To continue our critical enquiry, in this chapter, we examine how the management constructs an image of community and enforces ideals through symbolic means and repeated acts.

The Urban Commute

Commuting, life course, overtime, and gift-giving are examples of ways in which workers, social relations, and community are shaped symbolically. Company ideology is reinforced daily through routine aspects of daily life as a commuter. As shown in chapter 2, the commute is more than function; it has a transformative effect on the minds and bodies of workers, preparing them for another day at the office, while reinforcing the idea of the workplace as a community. The commute regulates the spacio-temporal experiences of the body.

"Commuter culture"—train culture and the cultural knowledge gained from commuting—defines urban life in Paul Noguchi's (1990) ethnography of the Japanese National Railway. He notes interesting observations of how commuters have "learned the technique of sleeping while standing as well as the best way to fold and read a newspaper to minimise the use of space" and enlists cultural metaphors of commuting. Noguchi comments on the role of train culture in regulating temporality in urban space in reference to literary and film narratives involving trains (pp. 41–8).

After the commute, in the lobby of the office building, polite greetings are in constant exchange. Workers often express gratitude for this haven of the office and the commute is frequent fodder for banter among employees in the morning. To my subjects the commute is an intimate feature of daily life. This practice that involves walking and being carried on trains and buses, unites the experience of workers as a group, through making commuters "independent of local roots" (de Certeau 1984: 111).

The commute, then, gives prior coherence to the experience that follows it, that of being situated in the office space, within a building, within respective departments and further in departmental subgroups with desks arranged as islands. The commute in a sense is also shared time by participating in a similar activity in an urban landscape. Time gives certainty. An activity that merely appears to be functional in nature is in fact a practice generating a sense of identification or belonging to a community.

Time, Work, Life Course

As much as the awareness of time is indispensable to the commute, so it is also to the experience of the workplace. In one session at the induction seminar for new recruits, we encountered time in linguistic expressions, in two cultural idioms that encapsulate the relation of life course (Brinton 1992; Plath 1980) to the experience of work and membership to a workplace community. First, working life is "a time of blossoming" (花開く時期), a time of fulfillment and enrichment. The job is hard work, but it should be enjoyable, thus rewarding. Second, it is said that the working years amount to "the most substantial period" (一番充実した時期) in the life course of the individual. In this context, the term "substantial" (jūjitsu) has both quantitative and qualitative connotations: a senior manager of the personnel department draws a time line on the blackboard representing the individual's life course. Between birth and death, a sizeable span is allocated to the time spent as an office worker. An average male worker who enters the company promptly after graduating from a 4-year university is 23 years of age and, if he works the whole time until retirement at 60 years of age, he will have served the company 37 years. That is almost half of one's life course. The average female worker will serve the company between a range of 3–15 years. At JCars, the majority of women work on average for five to seven years, and many are choosing to prolong their careers (after marriage and postpregnancy) as I found on a follow-up visit in 2003. Once a woman retires from a large company it is generally difficult to return to a similar office job; although, it is possible to return to work at JCars, and many return as part-time workers (cf. Matsunaga 1995).

The effective use of time is vital, the manager states:

会社で働いている時間は自分の時間ではない。給料を貰っている限り会社にいる時間は会社の物なので、これは入社した時点からしっかり解ってもらいたい。会社にいる時間は少なくとも七時間四十五分で、その間はしっかり働いてもらいたい。時間のけじめをつけて一日を有効に使って欲しい、それには定時の五時四十分がやって来るのを待つのではない。又、会社にいる時間が面白くないから定時がやってくるのを待ち望み、定時後にしか生きがいを感じ求めているようではどこかがおかしいそれとも間違っている。社員としての自覚が足りないことだ。

[The time spent at work is not your own. In fact, in so far as you receive wages, the time you are at work belongs to the company, and I want you to firmly understand this from the moment you enter the company. At the very least, you work for 7 hours and 45 minutes, and for this duration I want you to work solidly. Be disciplined in your use of time and use the day efficiently. As for this, it doesn't mean waiting for the arrival of the end of the working day at 17:40. If you're waiting and wishing for the end of the work day because it's dull, and you only feel and seek a zest for life only in the after-hours, that means something is going wrong or your thinking is wrong. It means that you lack awareness of what it means to be a member of the company.]

As we can see, new recruits are inculcated with the responsibility of managing time in daily routines as well as its quality. While young workers are accorded personal agency with respect to time, time is nonetheless owned by the company, which management have rights to control. This mixed meaning of time is retained throughout one's career. Furthermore, as time is integral to the workplace community, we can observe workers using time as a symbolic means to evaluate and qualify a good or ideal worker.

Daily Regulation of Time

Each morning an ambient electronic tune marks the onset of daily office time, once at 08:50 to remind workers to take their seats and organize their daily work schedule in their minds, and again at 09:00 when the official day begins. Most employees aim to arrive by 08:30, the reality is crowded lifts and chaos in the women's changing room around 08:45–08:50. By 08:55 there are an odd number of men and women slipping through the doors of the lift and into their seats, but tardiness is noted and remembered.[1] The third and fourth calls signal the beginning and end of the 55-minute lunch break at noon. All workers take their lunch during this time. There is no office canteen and lunch is eaten at desks in the main office or in adjacent meeting rooms. If workers go out to a restaurant, they are conscientious about returning on time; many workers are seen half-running back to the office. When eating in workers either get a take away from any of the shops in the shopping district (*shōtengai*), bring a homemade *o-bentō*, or order a *kyūshoku*.

A short five-minute break at 15:00 is allocated to *Rajio Taiso*: a tune plays out to accompany the set of movements designed to warm up the muscles and joints. The exercise is usually performed as a group activity in the early morning during school holidays in summer, and at the beginning of every PE class in schools. It is also broadcast daily on television. Workers are free to participate or refuse. No members of the personnel department take part, they seem embarrassed; but in the overseas department, the managers stand up and stretch out, and are often seen soliciting the participation of others nearby.

Overtime: Cultural Meanings

The end of the official workday arrives at 17:40. A day in the office totals 8 hours and 40 minutes (including the 1-hour lunch break). The majority of office workers don't leave immediately; usually they stay a further minimum of 20 minutes. We might say this is mini overtime. The reasons are altruistic, to purposefully convey courtesy, respect, and sympathy for those working overtime. On another level, however, this is an implicit rule of conduct. It is known that workers should not be eager to leave the office unless they have pressing obligations of a family or health oriented nature. This is captured by Capucine's experience. She interned at JCars for three months:

> The Japanese do lots of overtime. However, past 6PM people chat and drink tea. They seem to work in a friendly atmosphere. Through overtime it's possible to strengthen their solidarity as workers. This is very good. However, there are people who feel they ought to work overtime because others do. In France, people work overtime when they have lots of work. This is because French workers want to maximise time with their families. In France, the company doesn't decide when workers start and finish work. (Capucine n.d., c. 1997)

Most female workers stay for the mini overtime. When they have no further work of their own, they shuffle papers or visit colleagues in the department offering sweets or snacks. These treats are received by the department as gifts from other companies, or bought back by workers from their weekend excursions. This offering of gifts (*o-miyage*) and snacks (*oyatsu*) symbolically express caring (*ki-kubari*), important to the maintenance of social relationships and a sense of community. In this context of mini overtime, women take on the role of carer in relation to colleagues who will work proper overtime. After this obligatory 20 minutes, on the whole, female workers are first to leave the office. They leave the changing rooms in small groups, leaving soft echoes of laughter and scents of fresh perfume in their wake.

In the realm of proper extended overtime, dedicated women regularly stay on late, their managers notice, appreciate their efforts, and evaluate their performance accordingly; this leads to the assignment of work involving greater responsibility. These managers compare the women's attitudes to their male counterparts, assessing them favorably in comparison to men who are less committed. On several occasions, Sakamoto-*shunin* has come into the office at weekends to meet deadlines. In addition, women in the overseas department whose work involves managing work across geographically distributed sites frequently stay until 10PM.

Most men stay on later than women. General managers (*buchō* class) and above often work long hours because, on top of their own tasks, they incur the delays of junior workers who are slow to pass on their finished work. But younger men stay on until 7 or 8PM as well. But men too display practical and attitudinal differences to overtime. On a regular week some work a few hours overtime for three nights, or some stay late every evening, whereas some stay for the obligatory 20 minutes and leave the office around the same time as most women. The duration of time worked is a barometer by which the management judges a worker's dedication; it generates effects in that commitment is assessed and merit is awarded. Thus, time is loaded with the management's values. And yet, working quickly and efficiently is a mark of a good worker.

As we can see, overtime has multiple meanings and fulfills different purposes. In an atypical thus noticeable example, overtime is used for self-promotion. This particular section head (*kachō*) in the personnel department stays at his desk but does no work.[2] He cannot be seen leaving before his boss, thus he waits out that time, during which he talks prolifically (disturbing others' work). His loquaciousness *is* overtime work, but only to himself. His efforts, while real, are built on false foundations, because he constructs a reputation out of show. Yet to his colleague's dismay and surprise, he is successful at what he does. The head of department (*torishi-mari-yaku*) seems to like him. When he is lucky, his boss will extend him a drinks invitation enabling him to network further in what male office workers jokingly refer to as "office time."[3] This chance is what he waits for, it seems.

Overtime is many things: obligation, expression of caring toward colleagues, real work, and work that is self-interested (maintaining bonds between seniors and juniors can take a blatant form as we saw above or it might be less so). What appears as "doing nothing" (see Jack's observation, note 2 below) in relation to time is to miss the point that this nothing is really productive and directed toward long-term gains. Thus, overtime is both socially oriented action as well as individual goal oriented action.

Loyalty of Junior Workers

Miura-san, in the accounts department, who has been working for 1½ years, on one occasion, stayed on late waiting for his manager to finish. Time stretched on, only two of them were left in the building, and trains had stopped running. Miura supported his manager despite being told in passing that there is no need for him to stay. When I inferred through my question that I would have gone home, he answered:

> 女性は遅くまで残業しなくてもいいし、規定上してはいけないけど。男性は上司が働いていると先に帰りにくい 男社員は上司に対しての義理があるし、それに背くことをすればこの会社におれなくなるよ。仕事を真剣にとってないと思われるし出世できないよ。

[Women don't need to stay so late doing overtime, the employment law prohibits it in any case. For men it's difficult to leave before your boss if he's working, male workers have an obligation to their boss, and if you betray him it'll be difficult to remain at the company. The boss would think I'm not taking the job seriously and I wouldn't get promoted.]

Miura explains that the main method of promotion of junior workers in Japanese companies depends almost certainly on the status of his senior manager. Sakamoto-*shunin*, a female worker, mentions that the same principle applies to female workers as well; a woman with a progressive and understanding boss will be given more challenging and important assignments. Thus, as promotion depends on the quality of the relationship fostered between the manager and the junior in question, it is important for juniors to demonstrate loyalty to his manager, which confers at the same time an impression of commitment and ability toward his own job.[4] Enthusiasm for one's job is projected onto the wider frame of improving human relationships (*ningen-kankei*) between the junior and the manager.[5]

For loyal workers like Miura, overtime is, first, a tool used for maintaining good relationships with senior workers in the hierarchy and, second, it is simply what corporate culture expects. It is important for management to shape employees who will become the future core of the organization as they in turn will be entrusted to sustain the idea of community.

Kawabata-san's method of nurturing the bond between himself and his boss is by showing dependency:

上司を安心させるのにも色々な意見を持っていってあげる。これも自分の好きな時に言いにいける。 回数では計れないが、心がけとして行くようにしている。

[To give my boss support/reassurance I take him my opinions on various matters. And this I can do whenever I want. I can't measure it in frequency, only in the sense that I try to be conscientious.]

On the other hand, the most senior and respected woman in the office, Sakamoto-*shunin*, observes that men fear the consequences of speaking their minds to their bosses. They do not want to come across as defiant, since this is detrimental to promotion. Therefore, in the relationship between juniors and bosses, the balance between "empathy," "dependency," and speaking one's mind must be calculated with care. "Empathy" means maintenance of consensus whereas "dependency" focuses on attendance to others (Lebra 1976: 44–65).

Working as late as Miura had, as mentioned above, is a rare occurrence. However, there are other occasions when Miura's section worked overnight into morning, 24 hours. And during the week-long summer holiday during *o-bon* (festival of the dead) he was on 24hr call from the office. Furthermore, Miura often cancels arrangements with friends and it is impossible to plan a social calendar in advance given the priority he is expected to place on fulfilling work obligations. Thus, he places the company ahead of his personal life. Time is determined *for* him; this is not to say that he likes his position of limited agency or that life is somehow easy because he doesn't contest the dominant discourse. The nature of conformity masks the inner struggle. Moreover, at work, as in other situations, *ningen-kankei* (human relations) is difficult:

俺は会社の嫌な事は飲んだら忘れちゃう。嫌な人の事は気にしても仕方ないけど、気になるのは事実、病根のようなものと考えればいいよ。この人が嫌と思っても俺の生活や世界に入ってくるわけじゃないし、仕事上付き合わないといけないと考える。でも、忘れてしまっていると思ったり、忘れよう、考えないようにすると、やっぱり夢で嫌な仕事の事とかが出てくる。

[I forget about annoying things at work when I drink. It's pointless to preoccupy myself with disagreeable people, but in reality, I do. I suppose it's best to think of them as benign tumors. If I don't like a certain person, I just think, that person is just a colleague who I must interact with at work and there is no possibility of him/her entering into my life or world. But when I'm aware of thinking that I've forgotten about an incident or about an annoying person, or when I attempt to forget or avoid thinking about it, invariably, it reappears in my dreams.]

There is a struggle to balance work and private life, and Miura is engaged in the process of drawing this line, however fragmented and prone his thoughts are to the dictates of the dominant discourse. Miura's concept of time is dominated by the normative values surrounding the dedication of junior members to the company; he follows this religiously. For him there is no alternative or creative means by which to challenge this notion of time. Meanwhile, he possesses a strong notion of his own space and life, void of work commitments, but this he cannot realize. Clearly, it is not only the company notion of time he embraces with conscientious dedication, but the larger, albeit traditional, salary man cultural ethic. There are, of course, other junior men for whom work takes precedence over personal lives. Like Miura, they are among the elite core of the company.

Junior men are required to live in company dormitories for a number of years regardless of whether they are able to commute from their parents' homes.[6] They can stay a maximum of seven years. As stated in the previous chapter, by coresidence in company dorms, personal friendships are formed around various activity groups (Rohlen 1974: 221–34) because the amount of leisure time spent together overlaps with associations between workers while at work (Cole 1971). The personnel department believes that communal living encourages young men to learn to subsist for themselves independently of maternal affection and helps to foster intimate relations with co-workers. This enforced residential style cultivates dedicated workers through a constant reminder that they belong to the workplace community. Through the communal lifestyle, the workers' sense of self itself is being critically reformulated and learned.

Twenty years before Miura's time, dormitory life was different in some crucial respects. Muro-*kachō*, a section head in the overseas department, enjoyed living among peers. However, on some nights they received impromptu call-outs from the steward in charge of the assembly lines (the dormitory is located near one of the plants). He supported the line workers until 3 or 4AM. He then left for the office at 8AM. He says these practices have been phased out. In previous generations, then, younger male employees served the company as a flexible labor force, providing round-the-clock labor. Much more was physically demanded of male employees. The salary man of the 1970s and early 1980s was mentally as well as physically strong and agile and very tired.

Junior workers who share Miura's attitude and behavior are closest to the traditional model of the ideal employee, but there are alternative masculinities among salary men as we will examine below. Masculinity does not represent a "certain type of man, but, rather, a way that men position themselves through discursive practices" (Connell and Messerschmidt 2005: 841).

Western-Style Section Head

Tanizaki-*kachō*, a section head in the overseas department, entered JCars when his *senpai* from university was recruiting on campus. Tanizaki consistently leaves the office by 6PM to spend time with his family. He has two young children. He even built their American style home himself. In his late thirties, he experienced long-term assignments abroad in Asia, the United States, and the EU. He prefers the Western way of clearly distinguishing work from home. But his way involves passing on his work to women in his section, yet he claims the credit at the end. He also avoids taking telephone calls from colleagues in need of his assistance in order to keep his workload down. Nevertheless, his managers assess his contribution to the section and company favorably. This way of working is not particularly Western, nor is it a fair way to ensure that one finishes work at a set time. His work style angers female workers of his section. One worker who is thus encumbered comments that Tanizaki is deluded to think that he has assimilated Western values and to think his way of working is acceptable to others. Having built a reputation as a good employee over time, Tanizaki comes across as a good example of agency and creativity, while highlighting the fact that even traditional organizations can accommodate personal interpretations to the notion of time and work.

We can see that the management discourse is important to or is more effective in shaping the thoughts and actions of some workers more than others. Individual factors such as age and status, and nature and style of the department to which one belongs, interact to determine the complicity shown by the individual worker. Although the managerial discourse conceptualizes the company as community, there are intricate variations and subsections within this community, each with their own take on the management discourse.

Self-Maintenance and Embodiment

Kubo-san is in his early thirties in the IT department. He also draws a clear distinction between company time and personal time but Kubo's circumstances differ from Tanizaki's above. Kubo does not work more than was necessary or rather, in his words, "I won't sacrifice myself for the sake of work" (自分を犠牲にしてまで働かない), that is, not catering to the demands of the corporation. Most workers on his floor notice how Kubo is last to arrive every morning. Kubo thinks his department is relaxed about leaving times and overtime, and expresses sympathy for men working under heads of department with traditional ideas of work and commitment. Kubo knows his behavior is unconventional, but that is within the range of acceptability.

His attitude to work when placed in context makes sense. As a disabled worker, he explains that he fits a different level of management logic. Kubo's disability is imperceptible, but a heart operation in his early twenties classed him as an individual with a first-rate disability (classed by degrees of severity). Merging both management's and his own personal perspective, Kubo describes his status as outside the mainstream of the career and promotional hierarchy. This perception of marginal status derives from level of education and disability: not having a college degree, his qualifications include a high school diploma and a certificate in an IT course; and the fact that his employment at JCars is by chance, as a manager of the personnel department attended the recruitment fair held at the employment office for disabled people in Osaka City.

Kubo was a rebel during his high school years, skipping classes to race motorbikes with the local gang (*bōsōzoku*). He now cherishes his silver sports car. Kubo's father is a factory worker; therefore, he never expected to become a white-collar worker in a large corporation.[7] Kubo estimates his position in the company in par to the material process that got him in, a token gesture of the corporation abiding by laws and society, and he rules himself out of the running for promotion to high ranks. He appears content about his career-destiny. He clearly defines his career via the dominant discourse of the company that set aside different expectations for disabled workers; he is complicit in this way, and gives as little as is expected of him. However, if the management expected the same dedication from disabled employees, then Kubo's attitude would be subversive. In terms of the salary man's appearance Kubo definitely subverts the dominant discourse.

In many ways, Kubo is unlike his male colleagues. He is uninterested in promotion; dispassionate about marriage and children, although he has a long-term girlfriend from high school; and, in identity and behavior displays a pattern of embodiment quite distinct from the hegemonic corporate ideal. Investing many hours improving his self-presentation, mostly to maintain his self-esteem, he uses a sun-bed and works out every day at the gym. Kubo developed his interest in self-maintenance, after his operation left him pale and weak. He is among the more fashion-conscious of the male employees, interested in expensive designer brands, the consumption of which he justifies as a significant form of investment, and he colors his hair a reddish-brown, which, as far as I can observe, makes him the only man in the office to do so. He also experiments with colorful shirts in shades of pink and lilac. Other men wear white or subtle shades in pale colors. In JCars, generally, paying a great deal of attention to fashion is seen as an effeminate obsession, but women overall are impressed by his efforts, partly because it is subversive. It is rare for men in junior ranks to contest the management discourse, as they fear jeopardizing their promotional chances. Kubo is aware that his behavior is contentious, but nothing deters him. Other employees either admire or scorn his style. The latter camp says that he fails to take the social status of the salary man seriously, but the internal, emotional dynamics and personal history that lead Kubo to place such importance on his appearance is invisible, just like his disability.

Most workers in the office do not know of his disability. He keeps it to himself because he does not want differential treatment from other workers or their pity. However, Kubo volunteered his story because he knew of my interest in workplace micro politics; my concern with issues of gender and difference made him think that his unique story would interest me. The personnel department who are uncomfortable discussing disability issues were thus surprised to learn that Kubo had been speaking to me. Generally, disability is not discussed and explicitly avoided when possible.[8]

From the beginning of one's career, workers are aware of the concept of company time put forward by the management discourse. Although individual variations exist to this understanding, as seen in my accounts of Miura, Tanizaki, and Kubo, time is a crucial force, acting on individuals, reshaping them according to the ideals of the workplace community. The extent to which the management discourse is able to guide the actions of the worker depends on the workers' personal background and beliefs. Workers' beliefs and career goals inform the degree to which they are willing to concede to management and, equally, determines the methods they use to contest it. In particular, the implicit rule of staying an additional 20 minutes at the end of the official working day for women; and for men, staying for longer overtime, functions to foster a sense of community. Resentment for overtime is not openly expressed in the workplace, a wry shrug on a bad day perhaps, if even that. Perhaps this evinces the efficacy of the management discourse. The importance of time as an internal dynamic constituting the guiding principles of the meaning of community is underscored when we contrast Jack's view of overtime as "needless" to the meaning of practice; overtime is an expression of support and a way of conforming to the rules of the community. Moreover, we can infer that vertical relations between male colleagues are held together by how each orients himself in relation to time, whereas in the previous chapter we saw how vertical relationships between women were held together by a kin-based metaphor of "sisterhood."

Furthermore, how a worker positions himself vis-à-vis time and work, loyalty and career, therefore, intrinsically refers to masculinity. The idea of community shapes masculinity. This links to Romit Dasgupta's (2000) work on the performance of masculinity among salary men. Masako Ishii-Kuntz (2003) explores the enabling and constraining forces acting on the emergence of multiple masculinities among Japanese fathers. Her research suggests the importance of understanding masculinities in the context of social interaction, everyday practice, the state, institutions, and ideologies (p. 213). Furthermore, as we saw in the case of Kubo-san, who is a salary man but comes from a blue-collar family, when we consider his masculine identity we need to look wider still, at how aspects of disability (Gerschick and Miller 1994) and working-class status (Roberson 2003) shape masculinity. Like the masculinities it shapes, time is not singular, it is multiple.

Gift-Giving

O-miyage

Colleagues frequently offer each other *o-miyage* (souvenir/a distinctive gift brought back from one's travels) (Rupp 2003: 70–2). Both men and women bring back gifts upon visiting another city, typically over the weekend. They buy a large box of confectioneries that can be shared around the office. These gifts are usually perishable food items. Katherine Rupp's (2003) ethnography explains how in gift-giving "the giver shows the receiver and the people with whom the receiver interacts how much the giver values the relationship with the receiver" (p. 72).

On a Monday following her trip to Nagoya, which her colleagues knew about, Ono-san left her *o-miyage* at home. She is jittery all day, fearful of her colleagues. They might spread gossip about her rudeness and lack of ability to follow simple social graces. Thus, first thing in the morning, loudly, she announces to her colleagues through talking to me, how she "forgot" the gift at home, implying that she had not forgotten to buy it. Ono is conveying that the mistake is genuine and not a refusal to give. She chooses the morning to announce this as she anticipates that her co-workers are expecting the gift to be given out later that day. This is Ono's way of managing the group's expectations; her degree of concern points to its importance, she says in a mixture of Japanese and English:

> 同じ部の女の人達は "食べ物" に対してすごく執着心があるねん。 しかも "おみや げ" は絶対に買ってくるものと思っているから尚さら困ります。 めっちゃ気を遣い ます。
>
> [The women in my department have a great attachment to food; rather, their sole attention is focused on food. Moreover, *o-miyage* buying is absolutely imperative to them so it's all the more difficult for me. I feel extremely apprehensive.]
>
> I think if I didn't explain to them verbally that I forgot, they'll think I'm rude. So, I'll tell them personally at 3PM that I forgot it.

Ono's colleagues are not greedy and obsessive about food, as she claims. The issue of gift-giving happens to center around food,[9] since it is the commodity most often exchanged in the office. The sharing of food constitutes a discourse that confers social values and outlines social dynamics that define the inclusion and exclusion from groups, rivalries, and tensions. Social relationships in the workplace community are to an extent superficial, built on a slippery premise of membership to the company, which does not automatically constitute a basis for the formation of strong bonds. As workplace relationships are fragile, ritualistic events such as gift-giving constitute an important means of fostering a sense of community. The timing of gift-exchange therefore is seen to have significance and moreover, the failure to observe social conventions has consequences: Ono disrupted the aspect of certainty in the process of gift-giving and offended her co-workers.

At JCars the event of gift exchange highlights the fragility of social relations between workers. Gift-giving as a ritual activity does not function to strengthen emotional ties between workers, rather, it functions to strengthen *the form* of community among workers. The enactment of this ritualistic gift-giving generates between those involved an "obligation to reciprocate": (Mauss 1990). It is a moral obligation of the worker to bring back gifts for the workers in the department, although this is not explicitly stated by any means because it is central to Japanese culture. As such, gift-giving generates a "feeling of obligation" at a personal level (Testart 1998). Yet, as the event is held in view of everyone in the department, gift-giving acts as a "public sanction" (pp. 99–100) in which the character or the moral of the worker is questioned, if not performed properly. The characteristic of ritual gift-giving observed in the company fuses the traditional dichotomy between private "feelings" and the public nature of the ritual posed in traditional anthropological studies (Asad 1993: 72). Gift-giving, then, functions to shape the actions of the worker toward ways that foster a sense of community.

Senior managers bring back *o-miyage*, especially as they have many business opportunities to travel abroad. Yuko Ogasawara (1998: 141) notes that male bank managers give presents to female workers in order to "curry favor with women" because; ". . . it is men who more often fear women; men . . . are afraid to offend women." At JCars the environment seems more amiable between men and women; whereas women fear women. Men's gifts therefore symbolize a manager's appreciation for women's work, and they return with expensive Western cosmetics; commonly lipstick, small bottles of perfume, or accessories such as make-up pouches and lipstick cases, which women could select. The women assess the manager's credibility, and social sophistication based on the quality or expense of the gifts. In particular, if a good degree of selection and good taste has gone into the choice of gifts, the manager's status is elevated in the opinion of the women (also see Ogasawara 1998: 76, 80). Thus, the moment of gift-exchange is a context used by women to define the status of their male managers.

The ritual element of this practice and the functions of the ritual are essential to building a sense of community. People have to constantly rework their status positions and put them into practice, and this is the point about the ritualistic aspect of gift-giving, which also functions similarly to the regulation of time in the company (see Munn 1992: 109), as a practice that regulates the bodies of workers (Foucault 1977), but through a process that works at the level of affect between workers. Time is one of the guiding principles invested by the authority of the management. This leads workers to act in ways that reinforce the management's ideas of the community. Gift-giving also is a performance leading to the expression of certain emotions (even superficially) that are appropriate to the sense of community. Ritual is a discrete mechanism that generates a certain analyzable practice and it is therefore, at the same time, an object of analysis and analytical concept.

Thus, the operation of the management discourse is actualized through the use of time and gift-giving, as everyday events in terms of ritual, as rituals are "routine" symbolic actions "interpretable as standing for some further *verbally definable*, but tacit, event" (Asad 1993: 57, italics in original). Ritual forms not only function as a mechanism of control—through the embodiment of social beliefs (Douglas 1996; Durkheim 2001) by promoting social cohesion and equilibrium (Mauss 1990), and through the exercise of micro politics of power (Foucault 1977), or through the performance of internalized rules "inscribed" (as in a text) by a higher social order (Asad 1993: 62)—but, ritual also mediates between the demands of structure (hierarchy or management discourse) and creative "antistructure" (workers' practice) that are in dichotomous relation to each other (Turner 1969).[10] Rituals, according to this interpretation, function in many ways. My purpose is not to theorize ritual processes *per se*, but, in connection to hierarchy and *dōki* groups as discussed in the previous chapter, to suggest how various routines can be interpreted as constituting ritual, and thereby attribute to it a particular function in the context of the workplace community at an everyday level.

O-chūgen and O-seibo

O-chūgen and *o-seibo* are summer and winter gifts (food items) sent in July/August and December respectively. These offerings cement relationships between colleagues in vertical relation to each other because the logic of the seasonal gift is such that the sender expresses gratitude, while the reception of the gift prompts and obliges the receiver to give more. Rupp (2003: 34–50) describes how gratitude and hierarchy are intertwined. Female workers in their twenties send seasonal gifts to their managers, whereas men begin sending their bosses gifts after age 30–35. As Joy Hendry (1993b) has shown, the wives of senior managers organize the sending of gifts on their behalf. Women send gifts only to their direct supervisor; older males must send many gifts. This is because senior men with good careers owe and anticipate gratitude to a greater number of seniors who in the past and future play an influential role in raising the man's career.

My mother (wife of a high-ranking manager) explains the system. A short note written by the sender's wife usually accompanies the gift. In turn, the recipient's wife returns a postcard thanking the sender; some male recipients will also pen a line or two. If, following the receipt of the postcard, there is a gift sent from the senior manager's wife to the household of the junior manager, this gift is a hint (*anji*) that signifies an appropriate end to the giving. Thus, the senior determines when parties should suspend the pattern of mutually felt obligation. This usually comes about when the senior man is no longer as involved directly and intimately in the junior man's career. Thus, in place of words to articulate meaning, the return gift marks the moment when individuals are removed from the obligation symbolized by the gift.

Senpai no Ogori

Seniors take care of their juniors by taking them out to dinner or lunch. On these occasions, the senior always pays for the junior member—*senpai no ogori*. It is the senior's right and duty. What this occasion allows is an attempt to smooth hierarchical relations that are in reality complicated as we saw in the previous chapter. Georges Bataille (1989) was fascinated by Mauss's analysis of potlatch (means of circulating wealth). He observed that the laws of potlatch holds explanatory power for a great range of human behaviors (p. 69). We will draw a connection between *ogori* and potlatch in this context—the ideal potlatch, according to Bataille, is one without a return. He is interested in the excess of the gift. We give away and lose. For what end? The squander is meaningless if nothing is acquired via this loss:

> Gift-giving has the virtue of a surpassing of the subject who gives, but in exchange for the object given, the subject appropriates the surpassing: He regards his virtue, that which he has the capacity for, as an asset, as a *power* that he now possesses. (Bataille 1989: 69, italics in original)

The *ogori*, a social gesture, but also one of pride, this expenditure without a return (the junior never pays, or, if he insists, the senior allows him to contribute a token amount), is then about the senior redressing or "acquiring rank" and claiming "prestige" (pp. 71–2). In other words, excess and gift are linked to the affirmation and restoration of selfhood for the senior, but, not only in this process is rank and prestige acquired, it is self-attributed. Thus, there is a certain irony and inherent ambiguity: "[i]t is the stubborn determination to treat as a disposable and usable *thing* that whose essence is sacred" (p. 73, italics in original). The sense here is that our thinking betrays us.

Organized Events

Symbolic practices of group making, such as potato digging trips (Kondo 1990a); company parties (Rohlen 1974); and company tours to popular resorts (Turner 1995) are common to organizations. Likewise, at JCars, among departmental section members, women go out to birthday lunches to celebrate colleague's birthdays. It is the role of the female colleague closest to the one whose birthday it is to organize the venue and make an advanced booking. This intimate party is comprised of approximately five or six women who put aside differences and unite on these occasions. This too is a formal, bonding ritual, as some of the birthday lunches I attended were celebratory in form only: the conversations being awkward (see Li-san's comments noted earlier). Nevertheless, a sense of obligation is felt on both sides and a sense of community is fostered.

Departmental dinners take place frequently, particularly at the beginning and end of the year, and in March and April when members of the department arrive or leave. The male managers look forward to these occasions when they can chat to their female colleagues about general topics. Most women are wary of these events as they do not like performing the wifely/hostess type role of serving men beers and dishing food from the main plate onto men's smaller individual dishes. However, women look on these dinners as a free meal.

The reputation of a department is made and remade in relation to these dinners: the personnel department among others is known for their frugality and lack of taste for choosing the worst or tackiest drinking holes (*izakaya*) that suit men but not women. Yet in contrast, women in the overseas department look forward to these occasions as their managers are sophisticated and charming (*dandii*), as well as gentlemanly, a trait the women thought the managers had acquired during their postings overseas.

The method of planning an *enkai* (party) in the overseas department is outlined here. A form is passed around to group members who indicate which nights they are free. After the date is decided, the organizer telephones restaurants, inquiring about prices and asking them to fax menus for perusal. She consults her manager with this information, her manger in turn asks his manger. At the start, about 10 restaurants are considered but she whittles this down to 5. Then she passes around a list of the five places and members vote for the one they most prefer. Finally, the organizer emails everyone with the details and map of the venue. The funds for these dinners are allocated to each department by the company, the amount depending on the number of career track workers present and departmental performance. In addition, employees contribute monthly to the dinner kitty.

An American intern also noted how these dinners function to foster a sense of community:

> [P]ersonal relationships and working relationships are often one and the same. I was amazed at how well some of the sections got along outside of work, and I could see that camaraderie functioning in the workplace as well. In contrast, I worked in one group where the relations between the workers wasn't very good and I believe it affected the performance of the section. Therefore, whenever I had the chance to go out with my co-workers after-hours, I always did, as much to enjoy myself and make friends, as to contribute to the smooth functioning of the section. (Ben 13/01/97)

Summary: Fostering a JCars Community

Time is not regulated in the ways we have seen above for journalists at a national newspaper (the English edition) in Tokyo. At the newspaper, the features writers and editors are also engaged in a team effort, yet, as the journalists who had previous experience of clerical jobs in large firms said, social relations at the newspaper are surprisingly egalitarian for a Japanese firm as traditional and large as theirs (the basic form of hierarchical relations existed in both companies). This is attributable to the majority of staff being Western or returnees, and the shift-based nature of work. Journalists take lunch breaks independently of the larger group and younger members take cigarette breaks without fear of arousing their manager's wrath.

Contrasting the two offices, given that workers are capable of managing their own time, one could interpret the marking of time in JCars as having symbolic purposes. The general aspect of repetition imparted into the practices of workers at JCars through the regulation of time is, in effect, very ritual-like, where the routines divide and control workers' practices in the way rituals do, and time could be used in this way to govern the bodies of workers at JCars, because their work is not shift-based.

The regulation of daily time through electronic audio signals is a common feature of Japanese educational institutions. Therefore, the regulation of time is learnt early on in life as schools take over the role of the parent as the disciplinarian. By enforcing structured time on workers in this way the company takes over this role from schools, thereby coming to assume the role of parent as the worker assumes the role of the child. Placed within these familial links, the didactic discourse of obligation and loyalty that govern the use of time begin to make sense. Thus, one can understand the significance attached to time as a hegemonic discourse or structure imposed on the daily practice of workers. The structuring of time is a practice in itself and meanings that are unique to each institution are produced by linking practice and time (Foucault 1977). Thus, the structuring of time enables the company to regulate the workers' practices. It is therefore not the quality of interrelationships among members of the company being "family"-like (*ie*) or any inherent harmonious quality of personal relations within hierarchical arrangements that induce workers to behave in a certain way. In other words, it is the management's creation of a social order based on a notion of time that produces "bodily dispositions" for the workers to act in accordance to it (cf., Bourdieu 1990: 75; Bourdieu 1991: 89).

The structuring of time, then, is a way to nurture a sense of community. This management discourse is "mechanistically discrete" in its articulation of the ideal community. Time is a pervasive term in the office, not only used normatively, in reference to deadlines for a particular piece of work, but more importantly, as a construct capable of carrying and conveying emotional factors whereby managers solicit notions of obligation and loyalty while avoiding explicit and didactic references. It is also discrete because not only is the ideal formed at the top of the hierarchy disguised as it is filtered downward, but the managers are caught in it as well. This is because the ideals are encoded within the cultural "habitus" (Bourdieu 1977) of the company. In this way, the management discourse and workers' practices are not oppositional categories.

The operation of the management discourse is akin to the orchestra that brings together numerous individuals with their own style of performing that manages to perform flawlessly without the actions of the conductor, as famously put by Pierre Bourdieu (1977: 72). Similarly, Roland Barthes (2005) pictures the operation of ideology as discourse (*ideosphere*) in the image of hands holding together planks of wood in the shape of the crate, once the nails are hammered in the shape is constituted, making this process an "inside constructed from outside" (p. 86).

What we are seeing is that routines are endlessly repeated in the daily practice of workers in the office, and through imposing a conformity to time, what takes place is the reinstatement of the management discourse on an everyday basis. The necessity of repetition suggests the fragility of the management discourse. As the ideology is ingrained within structures of practice, without the repetitive practice, the ideology is redundant, which brings my argument back to the theory that ideology is embodied in repetition (Bourdieu 1977; 1991). Moreover, it seems a fair point to suggest that the repetitive process of signalling time is a way of making the otherwise fragile cultural mores of the office concrete. In other words, time is a way of shaping the habitus that is governed by the management discourse. Thus, without this repetition, management fears that the dominant culture of the organization and social relations within it would wither, although the observation of the newspaper office cited earlier in this section suggests that the management's fears at this company are unfounded. Yet it is not impossible, as, since the economic downturn, Japanese society is showing increasing signs of fragmentation.

Another point highlighted by my examples of Kubo and Tanizaki, whose practice is contrary to conventionally acceptable or expected behavior, is that the management discourse is not as strong as it appears. I have attempted to convey through their particular examples the existence of flexibility and agency in the way an individual might interpret the meanings of seemingly concrete ideologies. This makes clear my main point, which is that the management's idea of community must be reinforced through aspects of ritualistic practice involving repetition and the control of time because not all workers will behave and think in ways management finds desirable. Moreover, in their examples we glimpsed how multiple masculinities are formed in relation to the management discourse. Each of these masculinities formed around different concerns expressed in relation to time and overtime: loyalty; work-life balance; and physical disability.

Social relationships in the workplace community are to an extent superficial, built on a slippery premise of membership to the company, which in itself does not automatically constitute a basis for the formation of strong bonds. As social relationships in the office are fragile, ritualistic events such as gift-giving constitute an important means of fostering a sense of community. Member's relations to each other and individual's reputations are scrutinized when forgetting *o-miyage* dangerously slips into the meaning of a refusal to give. Departmental dinners and birthday lunch celebrations among women within departments all contribute to a sense of community. We also see that the hierarchical relationship is central to the giving of seasonal gifts and *ogori*. As shown by these situations of giving, expressing gratitude, incurring obligation, repairing relationships, gaining prestige, rank, and power are dynamics contained in the overall framework of community ideology.

The company is not an entity "unified by a community of interests" (Williams 1988: 75). Workers have various reasons and motives for being there, and their outlook of life varies in each case. These individual variations lead to varying degrees of identification and conformity to the management discourse. In this chapter we saw the styles in which the community is fostered, created, or "imagined" (Anderson 1991). The management discourse has multiple ways of imagining the workplace community and conveying it, and finds several mechanisms through which to inculcate their notions upon workers. The management discourse holds the community together; without the rituals that act to reinforce and articulate the values of the community on a daily basis, the community would not disintegrate, but, could perhaps deviate more from the ideal.

In the next chapter we will examine the spatial dimensions of the operation of the management discourse.

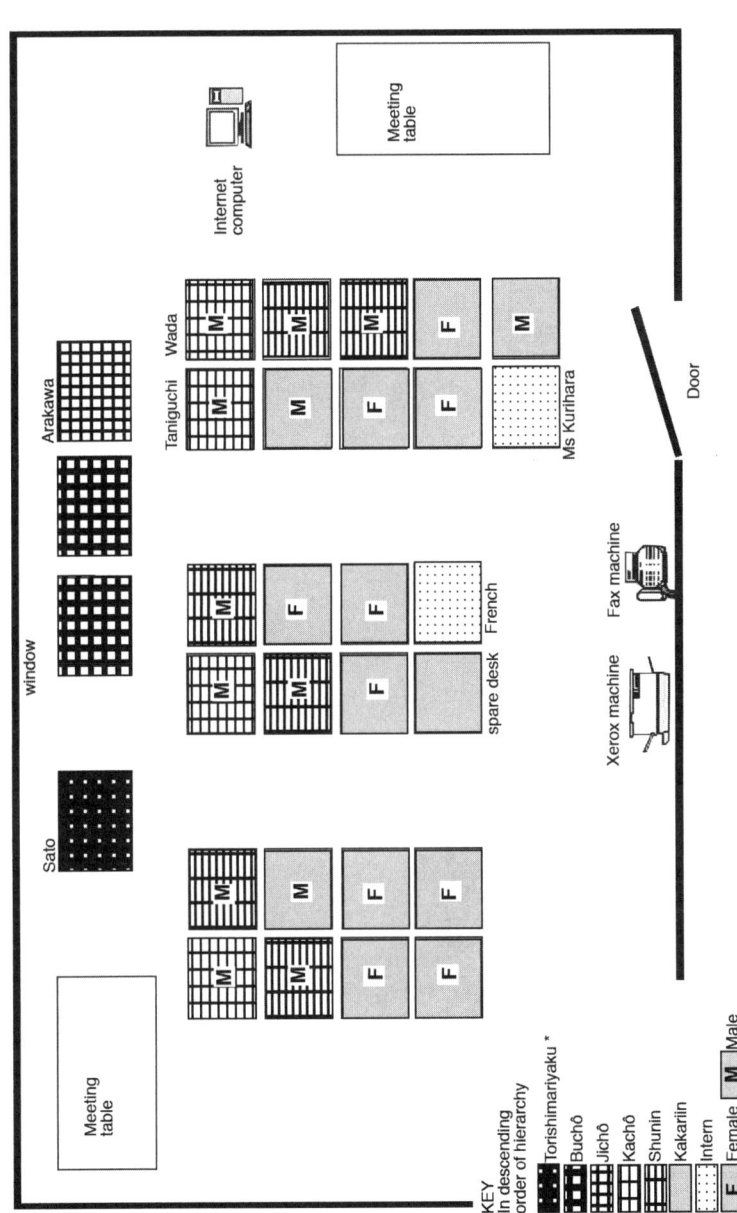

Plate 1 Personnel Department, January 1998 (pre-renovation floor plan/seating order)

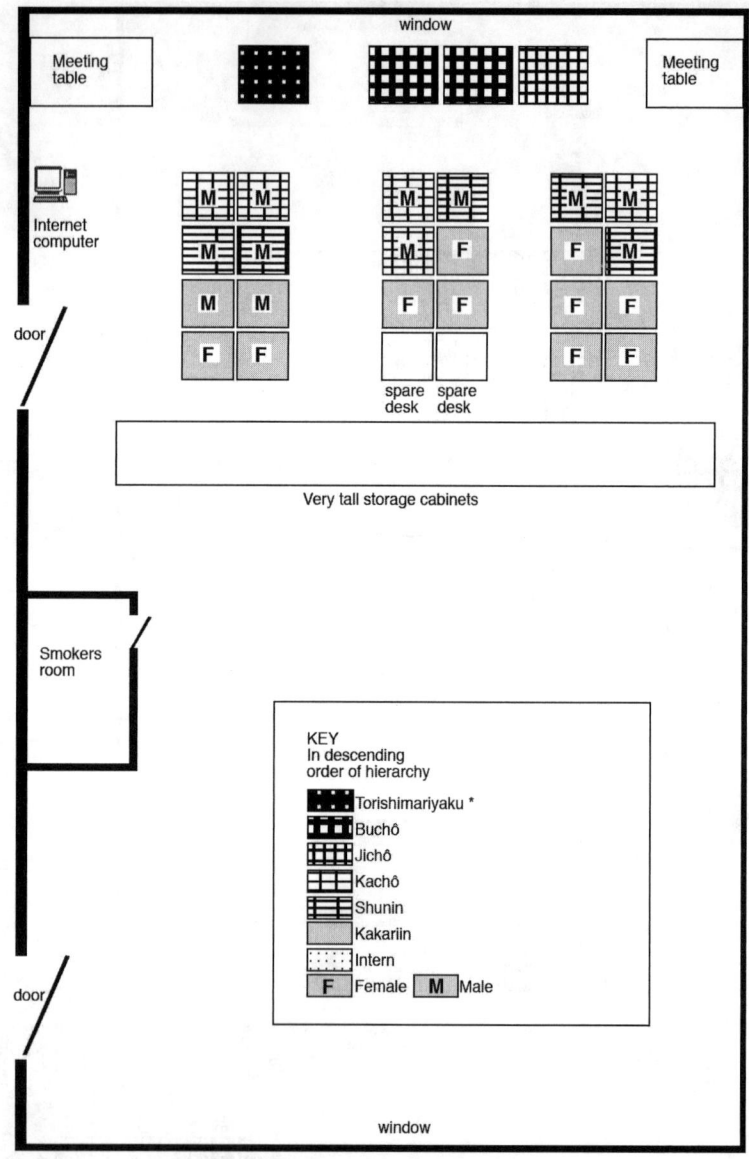

Plate 2 Personnel Department, January 1999 (post-renovation floor plan/seating order)

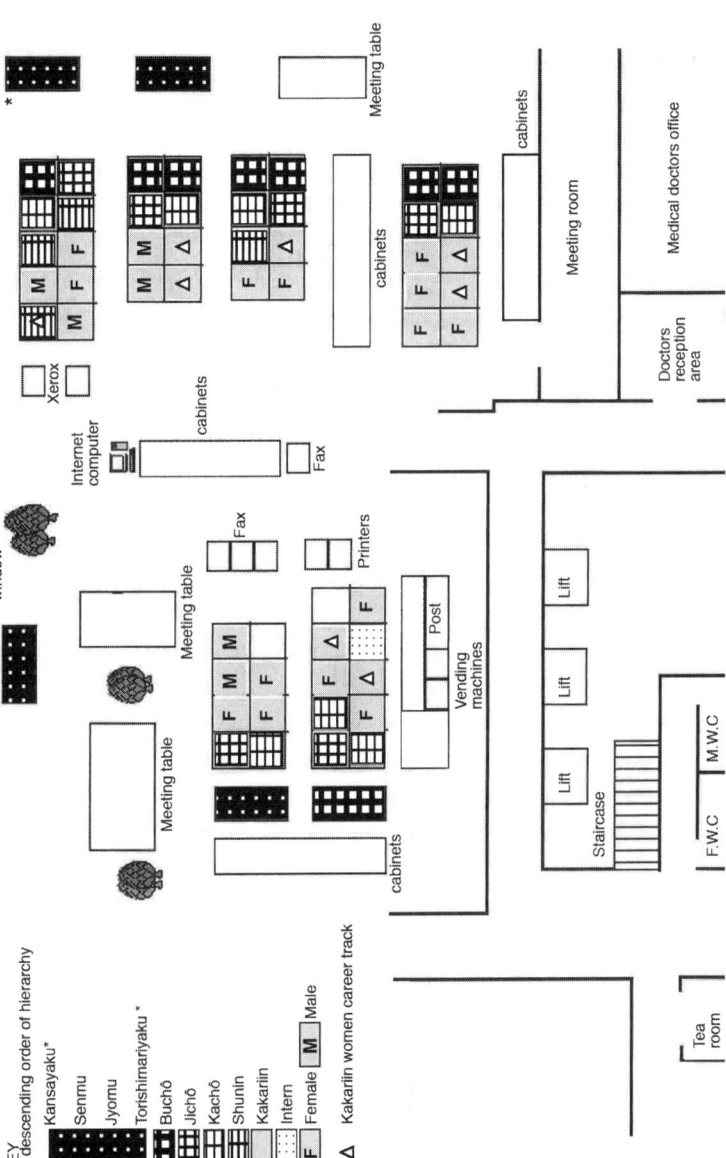

Plate 3 The Overseas Department, June 1998 (floor plan)

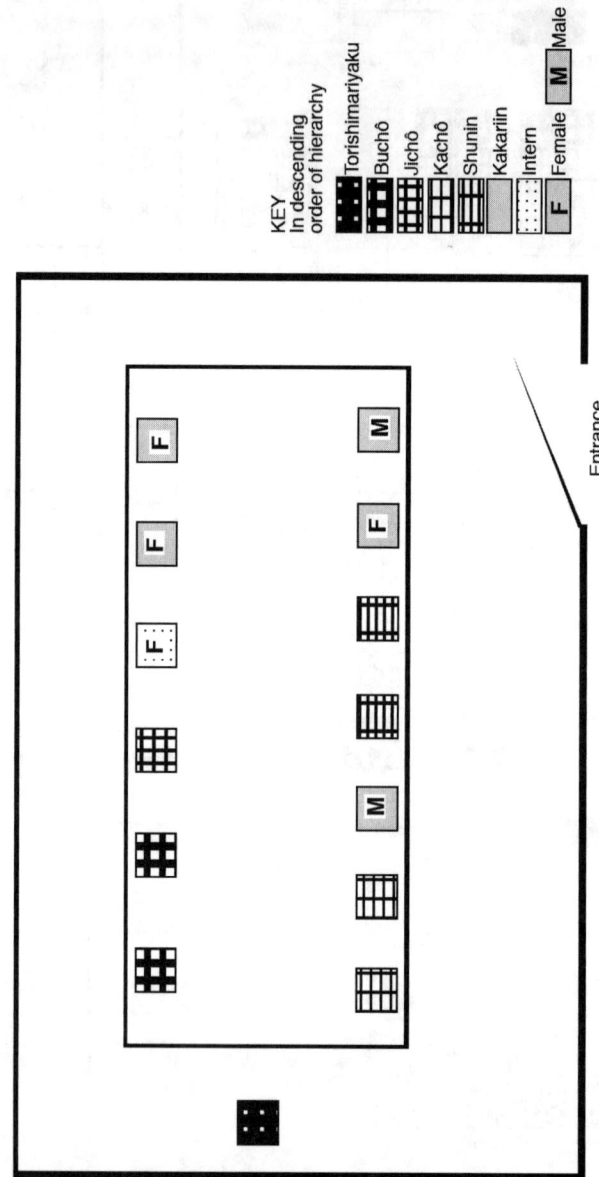

Plate 4 Seating Arrangements at an Office Party: Variations to hierarchy in different settings

CHAPTER 7

Spatial Practices and Hierarchy

In chapter 5, we saw the ways language presents the body in action, thus language being a mode of behavior (Butler 2004; Riley 2005). In effect, language is inseparable from the body; still, equally, it can be said that not all thought is expressed in language, but the unspoken knowledge can find expression through the body (Moore 1994). Our analysis in chapter 6 explored this latter premise in detail, and a picture of a workplace community shaped through a range of symbolic practices centered on time such as the commute, overtime, and gift-giving began to emerge. Thus, we wish to conceive of experience or knowledge in its widest sense: both linguistic and embodied. Allowing these two complimentary perspectives to steer us further into our remaining analysis, this chapter brings the spatial dimensions of office life to the forefront of our attention.[1] The more something appears to us to be natural and normal we neglect to question it; it recedes quietly and pales. I demonstrate that spatial practices in workplaces offer analytical purchase. Space acquires meaning through the practices it sustains (Bourdieu 1977; Moore 1986). And space is not ontologically given; it is discursively mapped and corporeally practiced (de Certeau 1984). Equally, space structures the type of practices that take place. Spaces are not only defined by geographies but by histories, therefore people's experience of space and situations are essentially through flows of daily experience (Ingold 1999). Similarly, experience is best described as routes taken through space rather than as references to fixed points (Clifford 1997).

We consider in relation to spatial practices the nature of social relations and symbolic practices of management that encourage all workers to maintain a sense of community. Our experience of space tells us something about the underlying structures of social organization, cultural values, relationships, power, and linguistic and bodily knowledge. Let us examine hierarchy in space, how it reproduces social organization (Durkheim 2001: 13–14; Giddens 1984); its characteristics—is it rigid, or is it fluid?;

and its role. This experience of hierarchy in space is considered from two intertwined perspectives: management and workers. Further, we look at the significance of gendered spaces.

I want to suggest that notions of hierarchy take on the *appearance* of the concrete when organized in the space of the office, but are in fact fragile structures that need to be constantly articulated and remade, thereby adopting numerous forms of representation. These representations of hierarchy in space deviate from the managerial ideal, in the sense that the hierarchical ordering of individuals in space (seating arrangements) does not translate accurately from the actual employment hierarchy. My subjects, who experience the incongruity between ideals and reality, construe hierarchy as an unstable organizational structure. As discussed in previous chapters, workers attempt to enact ideal forms of behavior to demonstrate their knowledge of social structure. With this in mind this chapter considers the creativity of the workers through the ways workers undermine patterns of conduct that are appropriate to vertical social relations. So, the model of hierarchy does not accurately represent the nature of social dynamics. How, then, do individual workers negotiate hierarchy within the space of the office, and what is the form and role of hierarchy in space?

Characteristics of Hierarchy[2]

Let us examine Plate 1. In the personnel department, according to the standard ordering of status by seniority/hierarchy, the managers seated along the window are of highest status (*jūyaku* and *kanri-shoku* ranks): facing the door relatively centrally sits the head of department, who is also a member of the board of directors (*torishimari-yaku*). Sato *torishimari-yaku* is the only one in the department to have a bronze plaque on the desk indicating name and rank. The others wear nametags bearing name, department (*busho*), and rank (*yaku-shoku*).[3] To Sato's left, in descending order of seniority, sits the general manager (*buchō*), vice manager (*jichō*), and the general section head (*kachō*).[4] Men seated at the head of each of the three sections are section heads (*kachō*), also called group leaders (*gurūpu-chō*). Next in order of seniority are the junior section leaders (*shunin*) who are men in their early thirties. They are followed by the women and men (*kakari-in*) who are not old enough to receive named posts. This seating pattern represents the hierarchy.

This section shows how hierarchy does not represent itself accurately in seating arrangements. This is because the actual allocation of status is inconsistent with the rules of promotion based on seniority by age as shown in table 2. The shortage of positions higher up in the hierarchy relative to the number of men eligible for promotion means that for

Table 2 Official Order of Departmental Hierarchy: Status Determined by Age

Age	Rank	Category
Fifties and over	Senior managing director (*senmu*) Managing director (*jōmu*) Director (*torishimari-yaku*)	Board of directors (*jūyaku*)
Forties	General manager (*buchō*)	Managerial staff (*kanri-shoku*)
Late thirties	Vice manager (*jichō*)	
Mid-thirties	General section head (*kachō*)	
Early thirties	Junior section leader (*shunin*)	Non-managerial staff (*jūgyō-in*)
Early twenties–late twenties	Workers (*kakari-in*)	

some men age will be inconsistent with status. For example, the older Arakawa-*jichō*, seated among the row of executives, is in a lower position than Tamori-*buchō* who is younger (the rank of *jichō* is one rank lower than *buchō*). As Arakawa-*jichō* is older, thereby, supposedly, more senior than Tamori-*buchō,* his rank indicates otherwise. Taniguchi-*kachō,* a section leader in the personnel department explained rather apologetically:

> …つまり一つの役職に対して人数が多いことなんです。　僕の世代の前に入社された人達の時は雇用状況も良く　終身雇用制度も関わり　同期の数も多かったんです。　しかし、このなかで出世するのも限られた人数で、役職名に限りがある代りに　席を与えることになっていました。　でも、また　これにも問題がありまして … 移動するにも部屋に限りがあって場所が無いんですよ。

> [… (I)n a manner of speaking, there are too many contenders for a single position. The generation of employees who entered the company before my time did so during good economic conditions, combined with the effects of the lifetime employment system, the *dōki* group was also large. However, a limited number of workers can be promoted, and to make up for the limited availability in status/rank positions we decided, in recompense, to allocate such people a (proper) place within the seating order. But, again, there are problems as well … the size of the room is limited so there is no space even for this move.]

Katori-san, a female worker who had been working for a year in this department describes, Arakawa-*jichō* as deserving of sympathy (*kawaisō*): "He wasn't promoted. Only his seat is among the eminent people" (彼は出世できなかったんだよ、席だけ偉い人のところにあるんだよ). This means that the traditional rule—status consistency with age—is symbolically preserved in the spatial arrangement of hierarchy. The symbolic representation compensates for actual status inconsistency with age.

Another anomalous case is this: Taniguchi-*kachō* and Wada-*kachō* are members of the same section, and of the same rank in terms of their seating position at the head of the section and role as section managers. Usually, there is one *kachō* to a section, but here there are two workers in the same section of equal official rank. Why are there two men of the same rank in this section?

Wada is often away visiting various professors in universities across Japan, building institutional human capital resource between JCars and universities. As his work is based outside of the office, when Wada is back at his desk, he has little to do. He rarely does overtime, unlike Taniguchi who frequently does. Taniguchi also goes on business trips, but less frequently than Wada. Women in the group say Taniguchi is hardworking, kind, and understanding as he never scolds or puts them down. He makes time for answering even the most basic of questions. Wada, on the other hand, recently joined the department from the sales department and the female workers in his section do not offer a flattering description of him.

As Wada is younger than Taniguchi by two years and an outsider transferred from a different department, the women are outraged that he is made the same rank as Taniguchi whom they respect. They speculate that Wada flattered his way (*gomasuri*) to Sato *torishimari-yaku* with his smooth talk.[5] The women fear that the *torishimari-yaku* cannot see through Wada's act— they genuinely respect and like Sato. They praise Taniguchi's true ability (*jitsuryoku*) to act as section leader and complete his own work, whereas they refer to Wada as a fake (*tatemae*).

Thus, these two men of equal rank have different roles as section leaders as they do personalities. Wada's role involves liaising with outside organizations, whereas Taniguchi oversees the workers in the section. This explains why there are two section heads in the same section. As such Wada should not have any authority over the workers of the section, as this is Taniguchi's role. However, as the status titles of these men are the same, Wada attempts to usurp Taniguchi's authority. This is because according to formal rules of hierarchy (status and rank) Wada ought to have/share Taniguchi's authority. The workers explain Wada's behavior by saying he is conscious of being seen by his section members as the outsider, given the nature of his work and status as newcomer. Wada emphasizes status by rank, not occupational role. However, the junior workers think Wada's blurring of rank and role is inappropriate and make their discontent known. The junior workers see need for only one section leader: Taniguchi.

The formal rule of hierarchy, as the company understands it, does not apply to the actual thinking of workers. Official hierarchies are guidelines, which appear concrete and logical, yet, as we have seen, the rules are not applied consistently in practice, nor is the premise fair; thus, hierarchy is difficult for junior workers to accept. To create a hierarchical order understandable and acceptable to them, junior workers call upon the assessment of personal characteristics and occupational roles of the two section leaders. This justifies the workers' ways of conceptualizing hierarchy in practice.

Rigid and Flexible Hierarchy

Relocation

Among women, the move (*hikkoshi*) from pre- to post-renovated offices generated much discussion about the use of space, whereas men saw it as no more than an ordinary, practical activity. Women expressed the following concerns about who has more or less space; is furthest from the eye of the bosses; and is closest to the path taken by people (*tsūro*). The move was seen as *the* opportunity to rearrange previous seating arrangements, either to move away from an undesirable individual, or to gain privacy.

In the overseas department where I worked at the time, during the renovation, small changes to the previous seating pattern were made on women's preferences. The managers are open to suggestions. In contrast, the personnel department, which moved to a newly renovated floor in December 1998, reproduced the original seating plan in the new space (cf., Plates 1 and 2). Unlike other departments there is no flexibility or choice for workers of the personnel department to negotiate where they can sit. It can be argued then, that the ordering of seats within this space is not only important as a visual map in which people can equate individuals with status and position, but that the principle of hierarchy *per se* has a symbolic significance to the personnel department. This is why hierarchy is a rigid concept for some departments whereas for others it is flexible. But why is this?

As shown in Plate 2, in the personnel department (post-renovation), the managers are located along the windows, and to recreate some form of wall-like separation from the other two departments located on the same floor, they erected ceiling to floor storage cabinets. This is a conscious effort to distinguish their department from others. In contrast, on the floor shared by the overseas department and the sales department, the divisions between them are in the form of linked series of low storage cabinets. One has a clear view of the entire floor.

The personnel department has good reason to cordon off its own space from other departments. The personnel department keeps files on all employees, decides transfers, promotion, and pay, and deals with welfare. Another role, the pedagogical instruction of employees, requires them to maintain and foster the official discourse of management within their own department. They set an example for other departments through seclusion. Their authority is derived in part from their central position in the management of the firm. Authority is reinforced by an image of elitism, fostered through a general sense of unapproachableness and silence in which the department is shrouded. In this way, the hierarchical structure is maintained rigidly and acquires symbolic significance in the personnel department.

Global Hierarchy

Moving on to another aspect of flexibility of hierarchy, on the whole, despite the yearly reorganization of workers to and from the head office to subsidiaries within Japan and abroad, workers are quick to observe these personnel changes, and adapt to the new positioning of workers within a social map. This also means that the display of hierarchy is not limited to the space within the head office. The vertical relation between the head office and subsidiary is manifested symbolically, and thereby company hierarchy maps on to a global space: when a Japanese manager visits a subsidiary, he is treated as someone a few ranks higher than his title suggests (Hamada 1992: 159). Simultaneously, the manager who is sent abroad embodies the reputation and performance of the particular subsidiary, which is dependent on the nature and success of projects that are developed there. Lower ranking managers who are less successful can boost their status and reputation by going to a good subsidiary. Alternatively, managers with no future can be sent to a domestic subsidiary with little clout. The general rule for the transfer of workers conforms to the description above (cf., Hamada 1992).

However, as the kind of projects developed and business performance of subsidiaries encounter constant flux, this hierarchy between different subsidiaries is likely to adjust accordingly. Status differences between subsidiaries also coincide with the images of the perceived status hierarchy for other societies: "East vs. West" and "centre vs. periphery" (Ben-Ari 1994). These symbols "travel across national boundaries" (ibid.) yet this Japan-centric view of status hierarchy for other societies is also subject to change, affected by the global economy and Japan's place within it, and by cultural notions about other countries that shift in accordance to tensions between the two (Ivy 1995). In this way, spaces are hierarchically interconnected; this then allows us to grasp differences *through* connection (Gupta and Ferguson 1992).

The acute sensitivity to a new knowledge of social maps among workers mirrors the fundamental importance of hierarchy to the notion of the company. At the same time, the practice of workers in the company offers competing discourses to the singular and deterministic view of social relations suggested by traditional notions of the workplace. While the official notion of hierarchy is explained as a rigid concept, subsequent meanings of hierarchy created by workers differ from the official discourse. In other words, hierarchy is given to a different production of meanings between the management and the workers, but the two interact. Grasping this conceptual distinction in the construal of hierarchy in space that differs between the management and workers, and even among the management, is necessary in order to understand the experience of work.

Office Parties

Hierarchy, when transposed to a relaxed social context, undergoes fluid transformation; this takes multiple forms. This section describes the parties of the personnel department unless otherwise stated.

In an office party situation (*enkai no ba*) the seating is ordered to reflect the positions of individuals as they are in the office (see Plate 4). At restaurants, companies book a room or a space sectioned off from the rest of the floor. The most senior member sits furthest from the door where the waiters move in and out, and surrounding him sit the other managers in descending order of rank. The women sit near the door. So does one junior man; he sits facing the door. His role (*kanji*) is to take care of the initial orders of drink and food, and further ordering of drinks that follow. He delivers the opening and closing speech for the *enkai*. As the *enkai* progresses the male managers move about in order to talk to people other than those nearest to them. Sometimes the women are asked to move closer to the senior managers, and if they refuse, the managers move themselves. Women and junior men pour drinks to the most senior and then to other managers before taking turns to serve among themselves.

In other *enkai* held by the overseas department, the seating is neatly ordered so that a man sits next to a woman. In this case rank does not matter so much, because managers wish to create a relaxed and informal atmosphere for interaction. However, the most senior member sits at the head of the table, which is again, furthest from the entrance. At *enkai* women serve food to men, which entails selecting and transferring items from the main dish onto smaller individual dishes. At *rishoku-parti* (standing party) women more than men move around within space, which is again because women take food to the men. Female workers do not like performing this hostess role but it is culturally expected of them; and it makes men feel appreciated.

The most senior manager leaves before the *niji-kai* (the second party). The *niji-kai* begins after the *enkai*, usually at a *kara-oke* bar (sing-along bar). The order of hierarchy in terms of seating does not matter here, but the men sit in one corner and the women in another. The manner of speech is informal although honorifics are still used.

The hierarchy experienced among workers thus takes multiple forms: based around a core shape of social organization as seen in the office seating pattern, at times loosely and at others quite precisely.

During work hours, workers in the personnel department are silent, but in this party environment they let their hair down, talking loudly and rapidly and asking personal questions, getting tactile, making hilarious jokes and observations—jokes are never too below the belt. The distance between people closes. For me, this difference in interactional style was very distinctive. Workers' individuality and personal style partied out of the closet. The stark difference in their behavior and language between social contexts surprised me: it is not that the difference should exist, but that it should be so marked. This is not the case in the overseas department: there is a definite party feel, people are more relaxed, less serious, people smile throughout, but, to me, the person remained constant.

Drink and Interaction

Various contexts of interaction give meaning to the space in which workers forge relationships (*ningen-kankei*). Thus, the meaning of space is equivalent to its context; this multiplies with a modicum of alcohol. Yuasa-*shunin* in the personnel department explains that sharing a drink with a boss can denote various circumstances of interaction (*o-sake no ba*). The context depends on the intersection of the nature of their relationship and the direction that the conversations might take (which is less predictable). He cites five contexts: (1) talking about ordinary topics and enjoying a drink together (普通の話をして飲むのを楽しむ); (2) giving advice to a junior worker (部下に対してアドバイスをする場); (3) discussing important business or settling an account (大切な話 - 商談を決める場); (4) grumbling to the boss (上司に悪口を言いに行く場); and (5) the boss complaining about his subordinates (上司が部下の悪口を溢しに行く場). As these examples of contexts for drinking show, the meaning of space is constituted through the nature of interaction it sustains, which varies between the casual and the formal. What Yuasa infers is that the meaning of space is defined by that particular moment; meaning within the relationship forms spontaneously.

Role of Hierarchy in the Open Plan Office

> . . . [S]patial practices in fact secretly structure the determining conditions of social life. (de Certeau 1984: 96)

Efficiency

In this section, we examine the role of hierarchy: efficiency; surveillance; and transparency. The open layout of this office is typical of most Japanese companies. The open plan office encourages efficient means of communication. It suits the system of passing on documents for approval; this takes place along lines of seating arranged hierarchically. It also facilitates keeping track of business knowledge. It is easy for all members of the section to gauge the precise nature of others' jobs and performance, as individual's conversations with section heads can be overheard. The management perspective on the open place office (which frames the workers' understanding) is promotion of efficiency and group cohesion by exposing the pace and content of information flow: efficiency and cohesion are mutually reinforcing. The interns from abroad, too, note the efficiency in the organization of Japanese offices. These interns have business and management studies degrees; they studied Japanese models of the workplace. Therefore efficiency might be emphasized in their accounts:

[In Japan everyone works together in a large room. Therefore, the flow of information is very good. For example, if someone is on the phone to a client, everyone can listen in.] (Capucine n.d., c. 1997)

Working in a Japanese office environment forced me to adjust to a whole new way of working. . . . I, of course, sat at the very end, being of the lowest status. Sitting next to and across from other desks without any walls or dividers, everybody's work could be constantly observed. It was extremely difficult to ignore telephone calls and personal conversations going on right beside me, but I eventually realised that perhaps this open layout allowed for better supervision and communication. If a group member is not doing their job, or if they have some personal problem, people instinctively notice, and a manager can step in and try and correct the problem. The downside of this communal layout is that individual productivity suffers because long periods of concentrated work become impossible with the constant distractions. (Ben 13/01/97)

[When looking at the difference between Japanese and French companies the first thing is that the working environment is very different. In France, workers have their own offices but in Japan everyone works in one large area. The merit of the Japanese system is ease of communication. The drawback is that one hears noises which have no relation to one's job, it is difficult to concentrate. Next, I mention two points about the Japanese system that does not exist in the French company. First, the method of signing approval; as everyone checks a piece of work the responsibility is dispersed. The drawback is that it is time-consuming. Next, the circular notices extend information to all, but requires multiple copies to be made.] (Arnaud 28/08/98)

Surveillance

The effective surveillance of workers by management is made possible through the seating arrangements in the open plan office. For example, senior members are seated around the outside forming a ring around the workers. In Plate 3, the positions of senior managers correspond to the darker colors, in contrast to the lighter colors showing the position of junior workers. Surveillance functions through this seating order organized into concentric rings, each more-or-less defined by descending order of rank. This arrangement of desks allows the "panoptic" managerial gaze to be directed from the outer periphery inward, toward the center where workers are seated. The "geometry" of the open plan office, an architecture that functions as a "disciplinary apparatus" is operational here (Foucault 1977: 173–5).[6]

For Foucault, the discourse of efficiency and the function of surveillance are synonymous (1977: 175): they are two characteristics of the dominant order. Hierarchy here suggests that placing workers under continual surveillance is a given right of management. Workers feel the pressure and regularly express an awareness of being subject to surveillance: they must work hard because managers are watching.

Hierarchies made visible in space facilitate communication and Japanese work methods but also function as a didactic tool for management to exercise control over the workers. Hierarchy is the primary means by which the management invests the space of the office with meaning.

Transparency

Efficiency and surveillance, managed by hierarchical structures, are vital to managerial authority. Transparency is another feature of design essential to the operation of hierarchy in the office: it enables hierarchy, imbued in authority, to be displayed effectively as a public "spectacle" or "theatrical ritual" (Foucault 1977).

Since board members work closely with the president of the company, the workers treat board members with high regard. Respect is manifested in the use of honorific language and in the physical distance that characterizes interaction between board members and junior management/workers. In addition to their role as board members they act as heads of department but sit in the office among the workers.

Miura-san avoids walking in front of the desk of a *torishimari-yaku* as much as possible, and takes long ways around in space. Investing a high degree of care in the practice of moving through the office is necessary in an open plan office, whereas it is unnecessary if board members have offices of their own. When Miura comes to the overseas department to speak with the *torishimari-yaku*, he stands at least three paces away from his desk, whereas I have seen Miura shoulder-to-shoulder in conversation with his *kachō*. Again this creation and maintenance of physical distance mirrors the positions of individuals within the social hierarchy.

Once or twice a handful of men including, *buchō* and *kachō,* received a fierce scolding by a board member regarding a blunder on a highly important task. The emotions of the *torishimari-yaku* resounded in an intemperate flow of Osaka dialect, which when used in this context resonates with considerable vehemence and power. This is a terrifying, awe-evoking spectacle, which engages workers on the entire floor. Afterward we all return to our tasks but the air has changed, the silence haunting the space in the aftermath of anger, and people are in fear and shock. I am struck by this spectacle for two reasons: first, by the demonstration of hierarchical distinctions, by its efficacy, that status difference is enacted so clearly in a powerful release of emotion, further implied through the language used; second, by the nature of the relationship that this act shows it to be, like that in the family between father and son. In this disciplinary context the senior member automatically assumes the position of the father. The significance of hierarchy as an organizing principal is reinforced among workers who witness this public display.

Another type of public display of hierarchy has a different meaning. Ono-san, a female worker in her sixth year at JCars and on the career track, is among the few junior workers who exchange informal conversations with the *jōmu* (managing director). I too was addressed by various *jōmu* on our floor, I think because they had affection for my father. Kimura-*jōmu* is talkative and a matchmaker. He offered to introduce his *Tōdai* (Tokyo University) graduate son to me, an employee of a prestigious trading company in Tokyo. He also wanted to introduce Ono-san to his son's friend for an *o-miai* (meeting with a view to marriage). He told Ono to bring a photograph and Curriculum Vitae, but he was unhappy with the photo she chose, the low-cut leopard print top was too informal. It is difficult to read the degree of in/formality of this situation.

The intimacy that transcends the extremes of hierarchical positions is a source of pride for Ono, and she speaks of it often; the attention indicates the senior's approval and acceptance of her character. She said this display of affection as though she is a daughter makes other women jealous, as the *jōmu* does not address them in similar manner. This theatrical or public display of affection by a senior member is effective in this context, because status is constructed in practice. For Ono, the public display is a necessary mechanism for legitimizing the construction of status. Intimacy is a basis for defining and confirming interpersonal ties. The hierarchical distance can be traversed in this instance as Ono's father as well as mine are well known to men of high rank. Their children also shared common experiences with both of us, having lived overseas as *kaigaishijo* (overseas Japanese children) and in Japan as *kikokushijo* (returnees). Intimacy and informality are more likely to take shape and to find forms of expression within social relations between workers who have family connections.

In contrast to the transparency of workers' movements in most areas of the office the space frequently occupied by the company president is opaque. The office of the president's support staff, reception rooms, and meeting rooms where he receives clients are located on the first floor. This floor is shrouded in mystique, as, unlike the other floors, there is little traffic of workers to it. Men and women who come to the first floor hurry through and whisper if they consult each other at all, even though they might be in the area of the lifts or stairs, far away from the inner sanctum of the director's haven, fitted as it is with rich burgundy colored carpets, oak panelled walls, and imposing dark brown leather sofas and armchairs. There is an immutable air of power and permanence. Tucked away from view of the main reception room and assistant's office, there is a full kitchen where meals and drinks for the president are prepared. The assistants, who are male and female, say this floor is the president's second home; he prefers to spend more of his time at work than at home.

A large majority of workers do not see the president. This way he remains an enigmatic figure (almost comparable with the emperor's relation to the Japanese people). What constitutes a worker's knowledge of the president is the existence of his private floor and through secondary knowledge via his essays printed in CN, CW, and in print media of various national and industrial newspapers. With regard to space, the eminence of the president increases through this distanced form of knowing. The status distinction between the president and the staff is embodied in the physical distance between them.

Gendered Spaces: Tea Room and Changing Room

Surveillance—Slipping Away

We now turn to gendered spaces. Let us remind ourselves of women's position in seating arrangements. As seen in Plates 1–4; in a standard interpretation of the social map, the head of the room (*kashira*) is associated with a position furthest from the door. This is where the most senior person in the room at the time is seated. In contrast, women sit by the door, near the boundary of the department. This lower position or liminal area indicates women's position in society (Lebra 1993) as well as their position within the hierarchy of the organization. How does this qualification play out in JCars? How do traditional symbols (i.e., seating arrangements) keep pace with changes in women's careers and occupational role that we discussed in chapter 4?

Women in the personnel department say that it is not because they are low status workers that they sit by the door. Rather, it is useful for women to be sitting closest to the door as they get up frequently to liaise with different departments (but this is because their responsibilities include menial tasks). Women collect the internal and external mail twice daily: in a rota system whereby two women go to the post room on the ground floor to collect and send the departmental post. The role includes sorting and delivering mail to the inbox of each member of the department.

Women sitting closest to the door invariably act as receptionists to redirect inquiries from workers wandering in from other departments. Nakahata-san, who sits closest to the door is frequently interrupted, which annoys her because visitors disrupt her flow of work. There were three females with the last name Yoshida in the department, and often visitors from other departments do not know their given names. Then Nakahata has to inquire about the details of the query at hand, in order to deduce from the nature of the topic exactly which Yoshida-san the visitor needs to see. Women are upset by disruptions to their work because they take their work seriously. This is because women do not perceive themselves to be of low status.

Yet, women can slip away. They can take advantage of their spatial position even though they dispute their low status by elevating it via occupational role. As women's seats are closest to the door, their movements go unhampered. In contrast, men do not have such freedom of movement (and, noticeably, do not attempt to create opportunities for freedom; however, smokers being exceptions). Thus, women frequently retreat to the tea room (*kyūtō-shitsu*) and changing rooms (*kōi-shitsu*) throughout the day. These are gendered spaces. It is a "tactic" (de Certeau 1984) employed by women to counter the "strategic" mechanism of management that aims to exert control over workers. For Foucault, "the body exists in space and must either submit to authority (through, for example, incarceration or surveillance in an organised space) or carve out particular spaces of resistance and freedom—'heterotopias'—from an otherwise repressive world" (Harvey 1989: 213). Women's responses to management surveillance are given below.

In the tea room some women keep make-up bags, hair products like wax and mousse, curling tongs, toothpaste, and toothbrushes. It is where some women eat illicit morning snacks or their breakfast bought on the way into work. Most women only stop in to make tea, but a minority of women use the tea room as a regular meeting place for chats and gossip. They organize their meeting beforehand by email to make effective use of the short time they manage to create. Sometimes women carry official envelopes, and, as they leave their desks, make sure the boss sees; it looks as though they are delivering a document to another floor. It legitimates their absence. This also buys time. As such, these women make use of the disciplining gaze of surveillance by which they are controlled. The tea room is a space from which they escape their managers

> いつも監視されてるからうっとうしいねん。 給湯室では監視されへんから気が抜けるねん。
>
> [It's annoying to be under constant surveillance. I can relax in the tea room because it's a place where I'm not under surveillance.]

Before the renovation, the tea room was larger, and more tucked away from view of the main corridor. Post-renovation, the door to the room was removed as well as the chairs. The room lost its comfortable feel. Some women say the changes are a management conspiracy because women are known to waste time in the tea rooms. But women have no means to substantiate their complaints. They are also enraged by placement of vending machines inside the tea room on some floors. They complain that men will intrude on their space because of this, and pity the women whose tea rooms had been changed in this way. They are *kawaisō* because they no longer have this space for themselves.

The appropriation of the tea room as a space for women proceeds naturally from what was once a woman's duty to serve tea. The logic is that it should be within women's rights to keep their space even though they no longer serve tea as a daily routine. This is a politicized discourse concerning the right for a private space. Discursive struggles over privacy occur because

> Both private and public places are heterogenous and not all space is clearly private or public. Space is thus subject to various territorializing and deterritorializing processes whereby local control is fixed, claimed, challenged, forfeited and privatised. (Duncan 1996: 129)

According to the geographer Ted Kilian (1998), public and private are not ontologically given categories; they are reified objects of analysis and thus are not characteristics of space; rather the distinction between public and private are expressions of power relations. Hence, both exist in every space (p. 115). (We wish to remember this point for our argument in chapter 8.) It is a matter of how "public" is defined, which then has bearing on the meaning of "private"; Kilian's example of a park can equally apply to the office:

> ... [P]ublic is defined by an idealized (and impossibly contradictory) vision of a space in which privacy-the ability to exclude-is not a necessary component of social relationships. Without the ability to exclude, the ability to limit contact, without boundaries, one is at the mercy of the power of others.
>
> ... "Privacy is viewed as the means of achieving individuality by providing the barriers necessary to enable the individual to make uncoerced choices

in life. Privacy could therefore be viewed as a mechanism for the realization of pluralism and tolerance." (Squires 1994, cited in text) (p. 125)

Returning to JCars following the renovation, we can see that the tea room is characterized by unstable/unclear boundaries, and hence contestation occurs. And these examples extend my argument so far that hierarchy is given over to different production of meanings.

Undecided Space

In theory, the tea room is equally open for use by men and women, although it is almost always women who use the facility. The room contains a sink and drying area with cupboards above it, an electric hob, a kettle, an array of cleaning products beneath the sink, and a grey, office cabinet of shoulder height. The door was removed during the renovation. This makes access by men easier as the room becomes less exclusive. I asked a *kachō* in my section why he does not make fresh tea in the tea room rather than buy from the vending machine. He says it's "complicated, time-consuming and bothersome" (ややこしいし、時間がかかるし面倒だから). I also put the question to a *kakari-in* in his mid-twenties, he said: "well, you know, it's because the tea room is a place where women hang out" (ええー、だって 給湯室は女の人の場所だもん).

The association of women with particular spaces through the course of usage is seen in Bourdieu's structural analysis of the Kabyle house, where the movement through a space constructs embodied knowledge, which thereby inscribes the space with associations of gender (Bourdieu 1977; 1990).

It is unclear to men and women whether the tea room is a reserve of female workers or if it is open to men and women. This undecided space mirrors both the historical continuum in the traditional equation of women with domestic tasks and the contemporary changes within women's jobs and status. Generally, the assumption is that the tea room is women's space.

Tea Making

Traditionally, female office workers made tea for other workers and visitors (Lo 1990; Ogasawara 1998; Pharr 1990; Roberts 1990). My mother worked in the overseas department at JCars during the early 1970s. She told me how the five women in her department formed groups of two and three. They served tea, twice daily, to the 12–13 men in the department. Such obligations pertaining solely to women reinforced domestic gender roles within the office. As this role hindered the equal treatment of women, tea making and serving was revoked by consensus (not by law) sometime in the 1980s, as a result of protests by numerous women. This was a common body of knowledge with which my subjects were familiar, although the change took place before their employment at JCars. It was part of a shared history of all female workers in the company. The protests were not organized movements (cf., Pharr 1990). Rather, they took the form of individual criticisms by women at one-on-one, annual appraisal meetings with their managers. They spoke out against a feudal company policy, which they believed kept down women's status. Thus, the consensus of women's voices reached the attention of male managers in a dispersed but unified way.

However, the female assistants to the *torishimari-yaku* continued to make tea at least three times a day, which they took to their desks. If a *torishimari-yaku* was receiving guests, it was the role of his assistant to serve tea. Other women no longer made tea for the men in their departments. The managers below the rank of *torishimari-yaku* either drank home-brewed tea brought in a thermos, or bought cans from the two vending machines located on each floor.

Cleaning Duties

Despite the procurement of a professional cleaning service, at the end of each working day, female workers clean the tea room. A cleaning rota (*sōji-tōban*) with approximately 40 women's names is drawn up and passed between departments on the same floor. The rota comes around to one individual about once every two months. No one knows why women clean. Is it a company directive or a simple voluntary gesture of recompense for the privilege of having a tea room? As the company hires cleaners, this practice can be part of a pedagogical, managerial discourse of discipline. Cleaning is routinely practiced in primary and secondary schools in an attempt to enforce certain, culturally specific, moral dispositions through embodiment. Furthermore, performing a standard task for one specific space would temporarily eliminate any sense of difference or hierarchy (in rank and by clerical and career tracks) between those sharing the facility. In this sense, the cleaning rota is a shared practice that generates a sense of unity and a common feeling of responsibility. All women feel this is a cumbersome chore, but acknowledge it as their responsibility as they are the exclusive users of the tea room.

After five months of fieldwork, I was included in this rota (I think through my "sister"). Although some women in my section including Ando-san and Mori-san thought I should be exempt, as I was an intern, my "sister" advised that it would be appropriate to perform the cleaning duty, as I was nevertheless, a member of the group in a loose sense. My "sister," Ono-san, said that those expressing the former opinion were doing so as *tatemae*, and their *hon-ne* was that I should clean as well. I took the diplomatic option.

On the bell marking the official end of the working day, as I have seen other women do, I wash the tea towels using detergent and replace them with the fresh, dry towels that had been washed the previous day. I then clean out the tea leaves, deposited throughout the day into a triangular container kept in the corner of the sink. Then I wash the viscous container along with the sink. I replace the bin liner, inside which I place a bit of old newspaper to soak up any extraneous liquid from the rubbish.

Exclusivity: By Gender

One afternoon as I made myself a cup of green tea, an unfamiliar face of a middle-aged worker appeared in the entrance to the tea room. As he drew closer, he posed a rhetorical statement: "Men aren't allowed in here, are they?" (男はここに入ったらあかんやろ). He then commented: "Well, it's not as though I'm going to assault you" (まー 襲うわけちゃうからな). He runs the cold tap and fills a *yu-nomi* (cup), gulps down the water, rinses the cup, and goes on his way. The underlying logic of the off-the-cuff comment registers an awareness that men cannot cross the line into women's spaces without their actions being labeled *sexu-hara* (sexual harassment). This man's mode of knowledge can be summarized through Henrietta Moore who draws on Pierre Bourdieu's phenomenological construal of body and knowledge:

> ...The material world that surrounds us is one in which we use our living bodies to give substance to the social distinctions and differences that under-pin social relations, symbolic systems, forms of labour, and quotidian intimacies.
>   ...Praxis is not simply about learning cultural rules by rote, it is about com-ing to an understanding of social distinctions through your body, and recognis-ing that your orientation in the world, your intellectual rationalisations, will always be based on that incorporated knowledge. (Moore 1994: 71–8)

His comment assumes knowledge of the "rules" of access regarding uses of the tea room, which further exemplifies how practice is constituted by historical or structural knowledge, and that this is reinforced through the everyday usage of space.

Exclusivity: Power among Women

We have seen how women's practice reinforces the space of the tea room as their own and how men are wary of crossing the boundary. Another aspect of the social organization of the tea room concerns the exercise of power among women; the contestation of status and reinforcement of hierarchy is enabled by discourses of rights to exclusive access.

Senior women like to claim the spaces of the tea room and changing room to themselves. The women who raise issue over exclusive rights to the use of space are not the most senior, but middle-ranking in terms of age. They have been employed at JCars between five and seven years. They use various indicators to signify their senior status over the juniors against whom they compete for the use of space. Status indicators include educational status; career or clerical tracks; popularity among co-workers; appearance and beauty. Discourses denoting differences among women evoke nominal or cultural notions of appropriate behavior.

These senior women complain when junior women use the changing room to fix their make-up at irregular times during the day. They say it is *their* privilege to be able to wander from their desks. Seniority accords them the right to escape the panoptic gaze. They object when they find juniors chatting in the changing room. A commotion occurred when one morning a new recruit was found lying down, spread out, comfortably asleep on her coat. If juniors sit on the benches the senior women complain among themselves. Notions of hierarchy based on year of entry or age become important during these times when women are in competition with each other. This is because senior women legitimize their privilege in the use of the changing room by notions of hierarchy borrowed from the dominant discourse.

Senior women are given larger lockers, and they complain if they are located too closely to a junior woman's. In this example, physical allocation of space should mirror the difference in the position of individuals in the social hierarchy. They also complain about the use of language by juniors: emphasizing the slack use of honorifics and lack of deferential behavior. In this, they often contrast their experiences as junior workers to the attitudes of the women who are currently in the most junior positions.

Junior women at times make tea for senior women, often their "sisters." The senior women return the favor but less frequently. In this, the senior women appropriate the same symbolic system of power about which they themselves object. Whereas women previously expressed deference to men through the act of serving tea, with the abolition of this custom, they make use of the system to establish power relations among themselves. (This is not always the case for more than half of all women.) Likewise, senior women use their supervisory role in assessing the state of the tea room to make explicit their status and power over junior women. If a *haken-shain* (temporary workers recruited through an agency) or junior member does a sloppy job of cleaning the tea room, it is used against them, and these women are subject to character assassination. The senior women are particularly harsh toward the *haken-shain*, they are considered to be insolent outsiders who wear too much make-up. Senior women appropriate mechanisms of surveillance in much the same way as it is used against them by male management.

Women of mid-status who are concerned to point out the inappropriate behavior of junior workers must believe in turn that their own status is in need of rearticulation. The constant redefining of status is indicative of the insecurities felt by senior women within the changing world of the office in a society veering toward fragmentation, where junior workers are less willing than their predecessors to conform to hierarchical notions of social relations.

Smokers

Smoking (Klein 1993) was permitted at the workers desks before the renovation. As only one floor in the building could accommodate a separate smoking room, on other floors it is still possible to smoke at one's desk even after the renovation, but no one does. The room for smokers was created on the floor shared by the personnel, accounting, and domestic sales departments (the floor layout is similar to the overseas department, Plate 3). The room accommodates a long sofa that seats about seven people. A large glass window looks onto the office space. The younger men prefer to sit at their desks to smoke as they can smoke more frequently, and they would not have to share the space with a senior manager, but they still go to the smoking room. The use of the smoking room mixes individuals of different ranks and this makes the juniors uncomfortable. The meanings of the word (*i-puku*) are (1) to have a smoke; and (2) to rest. Thus, being in the smoking room negates the second purpose if we assume the two meanings are connected.

Smoking in the office is associated with men. Women do not smoke anywhere in the office as this is seen as an undesirable habit particularly for any respectable middle-class female. However, they smoke in secret in spaces of the office building. As a cultural trend, young middle-class women in their early twenties are increasingly taking up smoking in public. At JCars, smoking is a means through which junior women contest class based models of behavior. To junior women in their early twenties, smoking is a cosmopolitan and rebellious activity; it allows them to contest the views of the older generation (from their perspective this category includes women in their late twenties).

Knowing the juniors' attitude, senior women who have been working in the office for approximately five years, describe the new recruits as supremely arrogant (*namaiki*), particularly as they are so bold as to take over one of the meeting rooms in the reception area. A group of them eats lunch there. This action is contrary to company rules as the use of meeting rooms should be avoided on the off-chance that company meetings with external visitors take place during the lunch hour.

These meeting rooms are attractive because they are secluded in complete privacy. Such seclusion is seen as a privilege of the senior women, particularly those who take over the tea room. It was revealed that some of these junior women were smoking in this meeting room during the lunch hour. It makes sense from a practical point of view. There is a large crystal ashtray in the middle of the table, and the room already smells of smoke. As mentioned above, smoking is associated with men in the office. As such, the smokers among the senior women did not attempt to smoke in the office. Senior female smokers had a crafty cigarette in the lavatories, but managers found them out and cautioned a few of them. Junior female smokers who are effective in tricking the management therefore annoyed senior women.

The changes witnessed among female employee attitudes, with regard to establishing careers (see chapter 4), as well as a shift in the conceptualization of the office as a space for both men and women is reflected in this issue of smoking. The offices of the national newspaper in Tokyo where I spent two weeks, is egalitarian, and there the women smoke with the men in the corridor outside their department. The attitude toward smoking among new female recruits at JCars seems to imply that they construe their position in a similar manner to female workers in the newspaper office as equal to men. The act of smoking in the office by junior women then, albeit in concealed but more dignified areas (compared to the lavatory), is an active contestation of gender roles where the traditional male-centered hierarchy and the hierarchy between females are contested. Smoking in the office is a symbolic practice among women that undermines hierarchy. Their behavior is also a mirror of the changing social world where more women have the freedom to smoke openly in public.

Refusal to Clean Desks and Ashtrays

Like serving tea to men, women once cleaned ashtrays. However, due to discussions with management they no longer have to do so. The complaints some women made against smoking were not simply to do with the unpleasantness of sitting next to a smoker; it is a contestation of gendered rules and male-dominated space. Further, following the renovation, a ban on smoking was issued on one floor. This suggests that the space of the office was construed less as a male space than as a space for both men and women. This refers to the management's perspective since they are active in establishing the smoking ban (except for in the smoking room). Women's refusal to clean ashtrays eradicated the explicit signification of the gendered division of labor. Through this practice, women redefined the meaning of space.

In addition to the tea room cleaning rota, in the morning before the working day begins, women also clean the desks of each member in their section. Men bow appreciatively at their desks and never react to this duty in a condescending way. Once for days I accidentally overlooked this duty.

As my "sister" continued to remind me, I finally cleaned the desks one day when workers were out at lunch. However, Muro-*kachō*, the manager sitting diagonally opposite me, to whom I spoke regularly as I ate at my desk, was there. He stopped me. He said cleaning desks was a feudal, gendered, and gratuitous activity. He said the workplace was everyone's space, no longer just male but female as well, and thus he could not bear the sight of me cleaning—a symbol of deference. The social sphere of the office, then, can be argued to be shifting, both conceptually and literally, to a space for both men and women.

Summary: Hierarchy and the Different Production of Meanings

This chapter gave an interpretation of the management's notion of space through the workers' perspectives on space; this entails production of different meanings.

Space is presented as synonymous with efficiency. In this regard, management espouses capitalist notions of space: which is infinitely open; desanctified—divorced from religious belief and emotion; assumes control over the distribution of workers; and employs mechanisms of surveillance (Marx 1967). The notion of space, from the perspective of management is similar to modern, continental notions of space, which is largely informed by rational and economic criteria. And, in fact, these are characteristics of ideology as we touched on in chapter 1.

The management discourse can persist and assume a notion of space with claims to rationality and objectivity, because the setting of the *open plan office* (*ōpun puran ofisu*) contains the assumption of a fair working environment. This is deduced from the connotations associated with the word "*ōpun*" that is borrowed from "open" in English usage: "exposed, direct, forthright, public, non-discriminatory, clear, undisguised, objective, unbiased" (The Concise Oxford Thesaurus 1997: 549). The loan word "open plan office" has been incorporated unaltered into Japanese daily usage (Kenkyusha's Japanese-English Dictionary 1974: 1311). The management discourse appropriates these meanings implicit to an open plan office, thereby articulating their idea in a mirror image of the object from which the meanings are borrowed.

Management assumes space as neutral and transparent. The space of the office is inscribed with their notion of power relations and decisions about the ranking of workers, which management expects the workers to sublimate as the normative view of social relations. In so far as hierarchical relations are presented as norms to be accepted, this suggests that management assumes space as an apolitical and unproblematic medium. The space of the office does not at first appear to be contrived, but clearly it is a medium that operates as a strategic mechanism not only of control but also to instill among workers a normative view of social relations. Conceptualizing space in this way enables management to shape workers according to their ideals. In other words, seating arrangements and the rule of hierarchy it idealizes, are another way through which management defines the "habitus" (Bourdieu 1977; 1990) of the office, insofar as both articulate structure. Although *habitus* is looser than an enforced ideology, I am drawing a parallel by comparing how these ideas gain acceptance, which gives way to conformity and orienting of practice in a specified way. Management expects workers to accept seating arrangements unquestioningly because it is an implicit method whereby actual social relations are disguised through symbolic presentation of ideal form.

Further, from the management's perspective, space assumes coherence. Such coherence was disputed by the workers' knowledge of space: space has multiple meanings, which alter in particular contexts. The relation of power between workers of different ranks did not go unquestioned by the workers. The promotional order, hence status order, is not as straightforward as the management discourse implies. The visual representation of status through seating arrangements is a structure permeated with contention and ambiguity. It is clear, then, that the management's notion of office space masks bias, contradictions, and inconsistencies inherent to managerial decisions. Thus hierarchies are not absolute but relatively arbitrary guidelines for social relations. In fact, the actual flexibility of hierarchy is masked by management, because it is vital to maintain authority by espousing ideas of utopian social relations. At best, seating arrangements provide a framework for workers to talk about and perceive social relations, but it does not determine these social relations.

In our discussion of gendered spaces, the concept of space enabled us to account for the movement of workers and social change in diachronic fashion. In fact, the discussion of space and hierarchy, and private and public, were emotive topics. Hierarchy permeates gendered spaces as well. In the tea room and changing room, senior women apply management's notions of hierarchy to legitimize their senior status over junior workers who are unwilling to conform to their norms. Also witnessed is the change in the conceptualization of the office as a space for men to a sense of shared space between men and women: the provision of a room for smokers; junior women smoking in a space where older cohorts of women would not have done; and a male manager's reaction to my performance of desk cleaning duty. All these are testaments to this change.

In the next chapter we also examine space and hierarchy but in the format of the company email network and office Intranet. Email and bulletin board is another space for the formation of community (this theme carries over from chapter 6 as well). Through the use of email, workers confront the notions of public and private space (in the nature of email and also in terms of engaging in personal activity within company time), and this relates to how we view the issue of the location of self to others as members within a bounded community. Through email use, workers make use of the dominant structure with minimal disruption to implicit rules of conduct contained within the principle of hierarchy. We will see how the company email network deepens bonds between individuals and, further, allows for expansion in the range of existing forms of social relationships. Friendships and romantic relationships will be considered in the context of vertical and horizontal ties mentioned in chapter 4.

CHAPTER 8

Spatial Practices of Mēru and Bulletin Board

The space of email and the bulletin board on JCars' corporate Intranet is a further dimension through which we explore the interconnectedness of workers in relation to hierarchy and community. Ethnographies of Internet use in Japanese white-collar organizations are not commonplace; as such, this book makes a significant contribution to an emerging field that studies new technologies in the context of local knowledge. Interactive communication transforms communicative styles and social structure but these are embedded in existing practices and power relations. Moreover, the transformation enables us to gain a different perspective on the idea of community from that which we considered in previous chapters.[1]

The following questions guide us through this chapter: what does management think about this alternative space and how does it differ from the way workers make use of it? What happens to hierarchy in this space, and how similar or different is it to hierarchy in the space of the open plan office? How does social interaction in this space coincide with or negate the notion of community that management attempts to foster?

Email technology is not new, but the restructuring of office communication at JCars took place shortly before my fieldwork in 1998. The timing, comparatively late for an industrialized country, was due to Japan's tightly regulated telecommunications structure (Coates 2000).

Email facilitates communication between workers within the company, as well as with and between workers in subsidiaries, with external firms, and it is used for both work-related and personal interaction. The properties of email allow this practice to be studied. Let us first consider the embodied nature of information in interactive communication:

情報とは、読んで字の如く、情　（なさけ）に　報（むく）いること…[こ]の表現
は…情報の持つ危険性への警告であ [る]。　情報は両刀の剣で、それを求める
者の身を誤らせもするものである。情報の本質は良き悪き とに別れる。良質の

情報は 広い範囲の人たちとの間に「情」のつながりを作っておいた元にある。 情報とは人と人の心のつながりの産物。

[When read as written, information means to requite feelings. This expression is a warning to the nature of the risk contained in information. Information is a two-edged sword; it can mislead or bring malice upon the one who seeks it. The essence of information is either good or bad. The genesis of information of good quality can be traced to the connections of "true feeling/emotion" established between a wide circle of acquaintances. Information is the product of the interconnectedness of peoples' hearts.] (CW 20/01/97)

情報社会は人と人との接触がなくなり、非人間的になると思うのは間違いです。

[It is fallacious to think that information society removes contact between people and becomes impersonal.] (Nikkei Business 20/12/99: 39)

The first quote from the JCars newsletter demonstrates the importance of information acquisition to business success. It also reads as a warning to workers not to lose sight of obligation, reciprocity, and the nurturing of human feeling in interaction despite the increasing pace of demand for information. It is a message, as in conjunction with the second quote, to consider information in anthropomorphic and emotive terms, and not in terms of antimonies. Further, this suggests that information exchanged via computers is very much like any other form of social exchange studied by anthropologists.

This is a study of spatial practices. The space made by the interface of two or more computers and the interaction of individuals at the end of two or more keyboards constitutes the space in which practices take place.[2] Importantly, the space is embedded in the everyday social world of the office. In addition to the visible network of social relations that take place in the office then, the company Local Area Network (LAN) system adds a further social space for interaction. This space is not divorced from the tangible field of interaction and knowledge of social maps that we explored in previous chapters. Rather, the social relations fostered through email and bulletin board communication are reinforced through the former and thus function in conjunction with it as well as in alternative and versatile ways.

Specifically, an email is a text.[3] The Japanese word used for email is the contracted form: *mēru*. The email text is a space through which workers build social relationships in a practice of exchange:

> [It is a] practiced place... the street geometrically defined by urban planning is transformed into a space by walkers. In the same way, an act of reading is the space produced by the practice of a particular place: a written text, i.e., a place constituted by a system of signs. (de Certeau 1984: 117)

Like conventional letters, email leaves traces on computer hard drives, and a message once sent cannot be retrieved. Email differs from the conventional letter in the speed of delivery and hence of communication. This allows for the possibility of higher frequency of interaction and in certain cases facilitates greater intimacy between those involved in the exchange,[4] although this means unsolicited emails by members of the workplace are also received. Email's asynchronous character means messages can be written and read when it is convenient to the recipient. Further, the company bulletin board engenders a feeling of community by drawing in employees who have not met face-to-face.

Installation of IT Systems

We examine the effects of email on work methods and social structure from a business perspective. We can evidence tension in both workers' and management's reactions to changes introduced via email use. The company's LAN system[5] of email has been in operation since the summer of 1997.[6] This system interconnects workers within the head office, and between domestic and global subsidiaries. On the LAN, workers can send email to any other account holder worldwide. However, the LAN does not offer actual Internet access; this is only available through a PC (one for each department) designated for this purpose.[7] Providing each worker with individual laptop computers connected to the LAN introduces several changes to work methods, and further influences forms of personal communication and dynamics of social relationships.

Prior to the installation of this integrated system in 1997, one desktop computer was shared between two workers in each section. Documents were created on word processing software and saved onto floppy disks. The main methods of sending documents to other departments was by telex, fax, and telephone where, once received, they were retyped from scratch if the receiver wished to alter the original text. With email, documents are sent by file attachments thus eliminating this double chore. LAN "provide simultaneous access to a text by more than one computer. In these cases the process of collective authorship, a practice that is common in some disciplines, is facilitated" (Poster 1990: 114).

Clearly, LAN increased the productivity of workers: (1) Female workers previously commissioned to photocopy circulars (dailies and weeklies) said distribution by email saved time and effort that enabled them to progress onto other tasks; (2) After a junior member finishes an important document, it goes through a rigorous screening procedure requiring the signatures of a senior manager and board member in charge of the department. This process often incurred delays when using departmental post. Emailing documents to the managers instead shortened the process by a day.

On the other hand, email and work documents sent via email attachments increase job inefficiency because documents are easier to send. Workers liberally send carbon copy attachments (CC:) to whomever they think might benefit by reading them. Thus, documents also reach workers with peripheral links to the matter at hand. Through reading, they might gain extra knowledge of others' job areas but, equally, it burdens their own workload because the knowledge is not directly useful to their work. In some cases email messages are printed out before the boss reads them, and this routine is bothersome to the assistants of *torishimari-yaku,* as assistant's jobs remain as labor intensive as prior to the advent of the LAN.

Kurihara-*shuseki* (my father), a managing director of a subsidiary in North America in his early fifties, adjusts his daily work schedule in order to accommodate the increased workload added by email. He usually leaves home one hour earlier than in days without email. On average, over 100 messages accumulate overnight. These are sent from various departments in the head office as well as from overseas subsidiaries. With email, my father keeps pace with the everyday developments taking place in the ever-changing business environment in Japan, which is an advantage for him, as managers such as himself often feel disadvantaged due to the physical distance from the head office. However, like other workers, he feels burdened by the sheer number of emails he receives.

Most female workers' comments on the antisocial aspects of email and attached documents echo my father's view, in that the vast number of emails is counterproductive to a worker's productivity.[8] Most junior women delete some unopened messages, particularly if the titles and the address of the sender indicate content with indirect relevance to their work, although the more senior women do not delete unread emails as other women did. By contrast to the women of junior rank, my father does not selectively delete messages in the way just mentioned and carefully reads each message, although he is selective in his reading order, where some are read before others. Given the similarity between the accounts of senior women and my father, the way workers deal with work-related email seems to be directly related to rank rather than to gender.

For Tanizaki-*kachō,* a manager in the overseas department, sending documents by email is problematic due to its inconsistency with Japanese work methods.

My subjects explain that there is no set methodology for the way a task is performed, that is, the method forms organically in relation to the particular task. This enables management to micro manage junior workers in every aspect of work, even in simple tasks such as writing fax messages. However, this work method requires a consensus on the way that task should be performed. Thus, this highly monitored pedagogical style is time-consuming and inefficient for business.

From a junior worker's perspective, this circular style of working can be demoralizing: Mattieu, an intern at the head office (1998), expresses extreme enmity toward this process. In one example, his *senpai* provides the information for the fax message, which Mattieu is asked to write, but his *senpai* rewrites Mattieu's message before passing it onto his senior. The message is returned to Mattieu listing further information that the manager wants included. By this stage, an hour and a half has gone into composing this one message. Mattieu says this reflects precision and high standards, but, in addition, it is the company's way of preventing his integration into JCars. His assignment of tasks with low responsibility makes him feel as though managers judge him to lack integrity and intelligence. We can understand Mattieu's frustration as he is eager to perform well in the organization but, whereas the experience might be unique to him, it seems to be common to junior workers', this we might recall from chapter 5, in relation to returnees' experience of translating articles for the company newsletter and Kasuga's similar encounter on composing an important business letter.

For managers, a different matter concerns them. With email some tasks can be completed and forwarded to other departments without consistent rerouting through senior managers of one's department as shown by Mattieu's example above. Thus, Tanizaki worries that email fosters a careless work ethic, allowing mistakes to go unnoticed and uncorrected before reaching the head of department. Put differently, the new work style bypasses the strict order of hierarchy (see chapter 6: how seating arrangements organize work flow). This issue has specific relevance to Tanizaki, as middle management (*kachō*) will incur responsibility for mistakes. Moreover, there is the potential threat of displacement that middle managers face, as email creates a new norm in workflow that functions by bypassing intermediate positions within the hierarchy. NB. As we read in chapter 6 Tanizaki favors a Western style work-life balance, yet here we see that he prefers traditional Japanese work methods that depend on hierarchical principles of organization (discussed in chapter 7).

It follows, then, that new forms of hierarchy established through email has repercussions for the existing hierarchy in the space of the office.[9] New technology defines the inefficiencies of the current structure of the organization, and inadvertently undermines traditional notions of hierarchy.

Tanizaki juxtaposes tradition against the new. Thus current work practices become imbued with ambiguous values. This view mirrors the larger, cultural discourse among the postwar generation. A form of "modernist nostalgia" exists for the historical and cultural past (tradition) in the current age of late capitalism, characterized by discontinuity and disruption of cultural forms and representations (Ivy 1995: 10).[10] This modernist nostalgia and anxiety termed "discourses of the vanishing" is kept in existence through a reinforcement of "the sense of absence that motivates its desires" (ibid.). Marilyn Ivy's insightful analysis has bearing on the microcosm of the Japanese office. Tanizaki's comment is framed by this discourse; his invocation of vanishing hierarchy in turn fixes and consecrates the traditional organization (cf., Nakane 1970); thereby averting its potential loss as new forms of technology alter work processes. In this discursive process, the traditional model of the company is reproduced, reinforced, and authenticated.

However, the majority of both men and women in their twenties and thirties are thrilled by this medium of communication and concur on the point that usefulness of email and the LAN system outweighs its negative aspects.

The following examples situate the ethnography of JCars in a wider societal context. At other companies, a similar process of technological integration between subsidiaries and the head office, within the *keiretsu* system (systematized network of affiliated companies) of business organization, is taking place. Toyota Automobiles announced in the late 1990s that an intercompany network would link worker's personal computers to the main server by early 2000. Certain tasks performed by the personnel department, such as processing expenditure claims, wage and approval forms, surveys of employment conditions and welfare forms, would be dealt with online, creating a paperless office. The installation of this integrated system parallels the recent shift in wage structure following the "abolition" of the *nenkō-seido* (seniority wage and promotion system) whereby wages of both clerical and technical staff would reflect professional ability and qualifications. Toyota aims to promote transparency in its wage and promotional structure (*CKS* 11/08/98).

Suntory Group is linking 166 companies within the group (e.g., Suntory Foods and Suntory Shopping) in an Internet system to facilitate the flow and sharing of information at horizontal levels. This will support outsourcing and new project development (*NSS* 18/08/98).

Practice of Economic Diversion

We now consider emailing on company time, which, from the management's perspective, presupposes a division in practices between work-related and personal emailing. The approach I take to my subject's use of email collapses these separate spheres of activity since, both quantitatively and qualitatively, personal email use is as important to their social world as work-related email. That said, when addressing the management's perspective, distinctions between work-related and personal emailing become relevant. This is because so far as management are concerned, the personal is to be distinguished from professional conduct: all personal mobile phones are switched off in the office and left in the changing rooms in the case of women, and personal calls from husbands or wives of employees are seldom made to office phones, thus, this norm applies to emailing practices.

Writing personal emails at work is a type of activity that de Certeau terms the "practice of economic diversion" (1984: 27). It describes the creative practices of workers that make use of dominant structures. De Certeau's approach removes the distinction between work and non-work thus enabling a dual consideration of structure and practice. This makes possible an account of worker's practices from the worker's perspective, freeing it from an exclusively managerial frame, although the latter can be equally accommodated.

Looking at "the operational models of popular culture" as de Certeau writes, is to "...transform...the object of our study and the place from which to study it" (1984: 25):

> The operational models of popular culture...exist in the heart of the strong-holds of contemporary economy. Take, for example, what in France is called *la perruque*, "the wig." *La perruque* is the worker's own work disguised as work for his employer. It differs from pilfering in that nothing of material value is stolen. It differs from absenteeism in that the worker is officially on the job. *La perruque* may be as simple a matter as a secretary's writing a love letter on "company time" or as complex as a cabinetmaker's "borrowing" a lathe to make a piece of furniture for his living room. Under different names in different countries this phenomenon is becoming more and more general, even if managers penalise it or "turn a blind eye" on it in order not to know about it. Accused of stealing or turning material to his own ends and using the machines for his own profit, the worker who indulges in *la perruque* actually diverts time (not goods, since he only uses scraps) from the factory for work that is free, creative, and precisely not directed towards profit. In the very place where the machine he must serve reigns supreme, he cunningly finds pleasure in finding a way to create gratuitous products whose sole purpose is to signify his own capabilities through his *work* and to confirm his solidarity with the other workers or his family through *spending* time in this way. (de Certeau 1984: 25–6, italics in original, my underlinings)

Workers are highly attuned to the management's definitions of "work," from the induction seminar as new recruits, the monitoring within the tutelage system, the symbolic practices that encourage workers to perform as the ideal worker, and through the presence of managers in the open plan office (chapters 5–7). Regardless of the pressure to perform as ideal workers, they continue to devote time to what the management sees as non-work.[11]

Thus, there are two levels to be considered in the conceptualization of "work" as this differs between management and workers, but preferably not through the distinction between work and non-work. However, the way workers talk about "work" and "non-work" is neither consistent nor uniform. Some workers distinguish between work and non-work mirroring the management's discourse when it suits them, while going against it at other times, whereas conscientious workers define "work" in alignment to the management's views nearly all of the time.

Thus, as the management discourse filters into the worker's notions of work in an erratic manner, to fully comprehend the dynamics of office life, paradoxically, work and non-work distinctions are analytically necessary. Therefore, collapsing work and non-work categories is possible only in theory; nevertheless, we wish to bear in mind de Certeau's perspective, primarily to avoid framing data and analysis in the dominant discourse, since this risks trivializing the practice of the workers. I think it is sufficient to say that in theory, I maintain that the non-work activity of writing private email on company time constitutes work, mainly because these workers disguised their activity as "work."

In conversations and interviews, I asked both female and male workers to estimate the proportion of emails written throughout the day for both work-related and personal purposes. The women's proportion of work-related email ranged between 50 and 80 percent.[12] This figure suggests that women use email mainly for work purposes. For some women, at least those whom I monitored closely given my proximity to them (in terms of seating arrangements or as friends), these figures are generous estimates. For example, one senior woman in the overseas department deals with work-related emails in the morning and spends the rest of the day emailing friends inside and outside the company, yet she claims she spends just 20 percent of time emailing friends.

The quantity of work-related emails received by workers reflects occupational roles and positions in the departmental or corporate hierarchy. For example, Yosano-san, an assistant to two members of *torishimari-yaku* in the overseas department, receives a greater number of work-related emails (often needing urgent replies) than the senior woman in the same department (mentioned above) although Yosano is her junior. Hence, Yosano cannot afford the time to email friends as often as her senior colleague, but Yosano emails her friends during spare moments. Further factors contributing to the quantity of personal emailing, include differing workloads between departments, which is completed within different time frames, even though to a degree all departments are interdependent; the manager's watchful eye; and individual attitudes toward work and personal life.

Compared to female workers, men email less for personal reasons, as many do not have the time to instigate let alone respond to personal emails. From my conversations and interviews with men, I arrived at this statistic: the proportion of personal emails written by men makes up approximately 2–5 percent of email use. There is of course no way of guarding against the tendency of male employees to deliberately underreport their personal email use. However, the female employees' discontent with men's infrequent responses to their email communication would appear to indicate that the men's estimation is not entirely fictional. Furthermore, men's emails when of a personal nature are often concise and remote in tone, and thus men appear insensitive to female recipients. These traits are often reasons for women's complaints. Older managers are even less inclined to use email for personal purposes, partly to do with their lack of familiarity with technology.

Mēru: A Web of Activity

In the office, emailing is an intricate web of activity that spirals outward engaging an ever-wider circle of participants. As one illustration of this, a phenomenon that fascinated me ceaselessly was the *jinsei uranai* (a type of fortune telling based on one's life course) that circulated among the women for an entire month (see Appendix 3). As various women claimed to be the first recipient (via friends from outside the company), we might assume that this *uranai* was popular throughout Japan at the time.

The *uranai* is an HTML (Hypertext Mark-Up Language) document and one inserts the total number of strokes of the Chinese characters of one's family and given names; sex; date of birth; and blood type. This reading produces pronouncements of events for every year of one's life. Most years claim that "one is living happily," but major crises such as illness or accidents of loved ones and of oneself are predicted, including year of marriage, birth of children, and the precise date of one's death.

The *uranai* includes a personality reading that provides a written explanation of results shown on the circular line graph: it determines in percentage terms one's wealth, success, and creative potential. Most women find the readings interesting and rationalize it, mostly persuaded by the scientific appeal presented by the graph. For two weeks, they searched the Internet for similar *uranai* using the freestanding PC with an Internet connection. They printed out their readings while maintaining discretion, waiting for an opportune moment until no managers queued at the printer. The women are inordinately animated, and pass the *uranai* around via email to people they know (both friend and foe), who passed it on to others. The flow and spread of the *uranai* from person to person shows how the hierarchical structuring of social relations can be bypassed temporarily. This example exemplifies the general point made earlier, about the new possibilities given to social relations in the office through the practice of email. We now explore romance via email.

Romance

In the darkness, a digital image of a girl (G) and boy (B) pair of computers glows on stage. Separated in space, a melancholy computerized voice calls out, and the two converse:

G: Hello, are you there?
B: Yes, I am here, what are you doing?
G: Dreaming of you.
B: Where are you?
G: I am working but dreaming.
B: Of what?
G: Of you.
B: I want to see you.
G: But you can't.
B: Why?
G: Because I am afraid.
B: But I think I might love you.
G: My darling I love you too.
B: Then can't we be together?
G: It's impossible.
B: This is my dream.
G: But I can never hold you.
B: I need you, I want you.
G: I am no good for you.
B: I don't care.
G: But I am a machine.
   Love is more than a simple switch.
B: I want to hold you.
G: I want to take you in
   And wrap you around my ribbon cable.
B: Will you be by my side?
G: This binary world is no place for love.
B: I want to be part of your neuron-net.
G: I can only ever love you from a distance.
B: You came to me in my dreams
   And told me you would come one day.
G: My hub is buzzing.
B: Now you are here, I can hold you.
G: All this information can go to hell.
B: Will you send our data packets into the world?
G: Hold my interface, squeeze my mouse, stroke my keyboard.
B: And then I too will become carbon again
   And our ashes will mingle in the ether.
G: One day my chip will burn out.
B: I must go before I die of heartbreak.
G: I must back-up.
B: I must go. (Antirom: RGB Show, September 1999)[13]

Anthropomorphized, the computers above take on human lives, voice, and emotion, in which they need no human operators (Nicholas Ryan, personal communication 06/06/99). Antirom's performance expresses the relationship between technology and the social:

> Positioned on the line dividing subjectivity and objectivity, computer writing brings a modicum of ambiguity into the clear and distinct world represented in Cartesian dualism. Human being faces machine in a disquieting specular relation: in its immateriality the machine mimics the human being. The mirror effect of the computer doubles the subject of writing; the human recognises himself itself in the uncanny immateriality of the machine. (Poster 1990: 111–12)

The dialogue above can be read as a relay of email messages: in the office, the female workers often speak of romance and relationships conducted through email. Email creates opportunities for getting to know others. It is ground for developing the beginnings of a romantic relationship, which can be taken further in face-to-face dating, for example, take the following.

Using email Sakura-san builds on the impression the men form of her on their first face-to-face meeting. She meets men through *konpa* gatherings organized by friends or in restaurants or bars. By using email to bolster this initial positive image, between the time of the group meeting and individual dates, Sakura overcomes her feelings of shyness that can potentially distort her performance during the date. It is easier to convey or construct through the email text a precise impression (to a degree) that fits the kind of woman her date finds desirable. Sakura can think carefully before committing herself to writing, something she cannot do at length on dates. The message is easily rewritten, which is impossible face-to-face; it is awkward to rephrase or retract a statement when one is attempting to feign confidence. She often shows me her carefully drafted emails composed to her paramours. She then forwards her compositions to other friends in the office to solicit their opinion before sending the text. In this way, email is also a concrete way to involve friends in the (usually unobserved) course of courtship. This stage is crucial as in-depth analysis of the relationship invariably follows.

Sakura is beautiful, and it is imperative to demonstrate a good sense of humor if she is to be taken seriously by prospective boyfriends. She has a linguistic-cultural advantage that allows her to demonstrate a distinctive wit: she gracefully mixes English words with Japanese. Sakura and other women complain that men reply in short messages, as though they do not care enough to compose at length. Younger men say they are flattered by the attention women give them in emails, but think it cumbersome as they cannot reply as frequently or in depth. They also feel inept as (potential) partners for not responding in the ideal way.

Sakura got to know Arishima-san, a co-worker, at a wedding party of another office member in December 1998. Sakura and Arishima work on the same floor, but until then had only exchanged formal *aisatsu*. Arishima was transferred to the head office from a subsidiary earlier that year. Arishima is quite shy, so Sakura began emailing him throughout the day in an effort to develop some rapport. Email is a perfect medium for this, as she said it is awkward to ask him out for a coffee immediately, as he might be put off. Using the pretext of an interview she asked me to research what he thought of women emailing him. She then became bolder and asked him out for supper, and after six months of dating and communicating through email during office hours, she wrote by email to announce their happy engagement.

Being in the same office as Arishima is reassuring to her as she had a concrete explanation for why he did not reply immediately to her email: she can see that he was away from his desk. This relationship then seems to have succeeded among other reasons as email added another layer of communication to face-to-face interaction. For Sakura email romance in the office is thrilling as most workers are looking for sources of gossip. Since email is silent, unlike conversations on telephones that can be overheard, she had the upper hand.

Meguro-san is having a difficult relationship with a man who is unresponsive to emails. Finding a potential replacement is also looking hopeless. Reconciled with the knowledge of her situation, rather pragmatically, she says email makes rejection easier to handle. She compensates for her sadness by evoking an alternative interpretation of email communication: that email is impersonal compared to letters or face-to-face interaction, thereby concluding that her current negative outcome is inevitable. The significance of the nonreply can be played down. However, thoughts of him preoccupied her.

Meguro's unavailable partner kept her dangling with occasional humorous replies; it kept her in the relationship. Meguro's experience with email shows above all that interpretation of the email text does not necessarily depend on shared meanings between the sender and recipient. Further, the text is a way of keeping another individual at a distance but just within reach. This is mutual distantiation: Meguro prefers email to communicate serious emotional issues concerning their relationship. This way she avoids any direct experience of embarrassment created by intense emotions, for that has been displaced and woven into the text. Further, email shields her from the recipient; she cannot observe his reaction (cf., Bauman 2003).

Ogino-san emailed her thoughts on email romance:

送る前に読み直したり考え直せるから実際会うのとは全然違うよ。　○○君もメールだったら普通って感じだけど会うとあまり喋らない方だもん。私だってメールで

は〇〇君に嫌われたくないからって わがままとか文句とか否定的なこととか書い
たりしない方やけど 心の中では溜め込んでると思うもん。 メールで少しは相手の
事を知ってるわけだから 会話するとき話題を見つけるのには良いと思うよ。

[As you can reread and rethink the message before sending it, it's completely different to actually seeing the person. 〇〇-*kun* seems normal by
email, but when you meet him, he's the quiet type. I don't write selfish
things, complain, or write negative comments to 〇〇-*kun* in emails because
I don't want him to dislike me, but I think I harbor grudges in my heart.
Since you know a little bit about the person through the email exchange,
it's good for finding a topic of conversation when you talk.] (Ogino
19/11/00)

Ogino's view of the relation between writer and text, where the writer is
unlike the real person, contrasts to Sakura's opinion above. For Sakura the
augmented self conveyed via textual practice is authentic enough because
she achieves her desired outcome. Constructing the self through text frees
subjects to assume multiple and shifting identities albeit to different ends,
it is a goal-oriented practice. How does the imagined-self integrate into the
self-other relation when authenticity of the subject is suspended between
unstable self-other relations?

The contrasting views of self and other conveyed by Sakura and Ogino
illustrate how "the screen-object and the writing-subject merge into an
unsettling simulation of unity" (Poster 1990: 111) and alerts us to the question of whether, or the degree to which, sentiments or character of individuals conveyed through a textual medium (email) can be taken at face
value.

Despite differences between Ogino and Sakura's opinions, there is nonetheless a common factor: email is useful in communication with men, as,
through differing measures and degrees, it enables a presentation of an
ideal self that is difficult to achieve in face-to-face interaction. Yet, both
measure email communication against face-to-face interaction. Therefore,
new forms of interaction enabled by interactive technology are grounded in
conventional ways of communication.

Friendships

In large organizations, one foregoes the opportunity to speak with individuals let alone establish friendships with them, but with email, and in
the example of the *uranai* above, some of this inability to make contact
with others is alleviated. Certainly, the field of friendships between female
workers opened up with email. I focus on women's friendships, since in the
office context, relationships between women and men are classed as either
colleagues or romantic relationships; although at times relations between

female and male colleagues are labeled as friendships (platonic), this is rare. When men and women are "friends" it is expressed by the nuanced term *o-tomodachi*, the addition of the prefix *o-* denotes gender difference, and hints at something more, it connotes sexuality. Rarely are *o-tomodachi* platonic relations, although they can be in some contexts (and, outside the workplace, it indicates platonic relations as experienced among nursery and primary school children).

Doi-san had never spoken to Teraoka-san in the same department beyond exchanging greetings. They sat in different subsections. They struck up a friendship since Doi emailed Teraoka on a work-related matter. Friendship developed because email made it possible to articulate their personalities through informal language and composition of the message. Rapport exists in email; it forms the foundation of their friendship. Email is a method of sustained communication in the office environment where informal conversation is discouraged, thus it has the potential to deepen individual bonds. Other women echo their example.

Sakura-san knows Yokota-san and Meguro-san both equally well. Sakura considers the two women her best friends albeit in different contexts: with Yokota she has informal, but slightly selective and refined conversation, but Sakura keeps no secrets with Meguro and they often go to clubs and bars together after work. Yokota and Sakura have known of one another for five years, but only exchanged formal greetings as they are from different departments, and, as senior-junior in the vertical relationship, Sakura thought it is inappropriate to get to know Yokota.

In fact, Sakura came to know Yokota through her friendship with Meguro who is Yokota's *dōki*. Sakura's impression of Yokota was reserve and refinement, thus the boisterous Sakura had assumed Yokota would take a negative view of her. Sakura and Meguro both have long manicured nails, and wear heavy eye make-up and sexy clothing. Such women in Japan are categorized as loud (*hade*). Whereas Yokota is quite the opposite, her make-up is discreet, and, her clothing to Sakura and Meguro's taste, is expensive but boring. Yokota falls into the Japanese category of women from sheltered and rich backgrounds (*o-jōsama*). Although Sakura and Meguro are in different *dōki* groups, they became friends initially through a shared taste in clothes and because one of Sakura's section members is Meguro's *dōki* member.

Through email contact Yokota and Sakura eventually became closer friends than Sakura was with Meguro. Sakura discovered Yokota's sense of humor and felt they had common sensibilities. Sakura says:

> I'm fashionable and not scared to be seen as not conforming to the dress code at work [clothes worn for the commute]. In this sense I'm more like Meguro. Although Yokota wants to be seen by people at work as an *o-jōsama*; gentle

and well-behaved (*otonashii*), deep down we share some kind of understanding. It's not imperative that I address Yokota using honorific form of address in our emails. Email made it possible to find out what she's really like. I discovered that Yokota is a lot cooler and less boring and prissy than my impression of her had been for all these years. My friendship with Yokota is important to me as I can relate to her intellectually, more than I can with Meguro. After all, Meguro attended a two-year women's college whereas Yokota like myself went to one of the prestigious four-year universities. But I like Meguro because she is kind hearted (*ninjō ga aru*). I used to think Meguro was my only true friend here in the office, but now I know I can count on Yokota's friendship just as much.

Some workers think that email does not intensify friendships if the friendship formed prior to the installation of LAN email. In addition, if friendship doesn't exist to begin with (i.e., two individuals in a vertical relationship) and if the social bond is based on an enforced association and not choice, for the communication to then become dependent on email means that the two workers speak less in face-to-face communication.

Hori-san uses email as a tool to subvert hierarchical relations; she can gossip or complain to other women about an annoying manager while he sits unawares within hearing distance. In addition, it is useful to be able to thank a manager for taking a group of them to supper, particularly in response to exclusive invitations where otherwise, some members of the department may feel jealous if they overheard her thanking him out loud.

Pubic and Private in Spatial Practice

Demarcation through Embodied Practices

In the previous chapter, we raised the point that public and private spaces are expressions of power relations, and coexist in every space. We can see how privacy is constructed through emailing practices, particularly for personal purposes. Let us consider the ideas of private/public through the behavior of my subjects:

Most workers interpret email as constituting a private act negotiated in a public space, particularly when the content of the email is personal. This fact is clear, initially, from the point of conscious misuse of company time. Moreover, the careful planning invested in disguising the activity from others is indicative of the nature of the act: engaging in a private act in the open plan office. Workers take to this type of behavior as a reaction to organizations that seek to control the workers' use of time and space. It is easily practiced by workers as this new medium dilutes the dictates of the managerial discourse. The fact that workers embody their knowledge of distinctions between notions of public/private and work-related/non-work related activities indexes their astute awareness of the official discourse.

At a glance, it is possible to identify personal email, as often women select various cute or unusual fonts or colors when writing private messages. Adding a personal touch to emails is an important means of self-expression. Female workers use creative means to personalize email texts: (1) Writing in Osaka dialect rather than in the standard Tokyo dialect, which is the form of written Japanese used in work-related emails; (2) To convey emotions women use extremely large font sizes often in brilliant shades of pink and other pastel colors, these being feminine colors.

Thus, individuality is rescued from depersonalized texts of computer writing (cf., Poster 1990: 113). Importantly, in personal emails, women conduct themselves in styles considered inappropriate to the office, and as such, this space created by email becomes essential to women.

In a similar way to women personalizing the email text, women use physical objects on their desks or on their person to embody personal taste or personalities. This creative outlet serves as an alternative to dress, as women wear uniforms. Most women carry pens and mechanical pencils in the top left hand pocket of their waistcoats: not the ordinary *Bic* variety but cute (*kawaii*), elaborate, colorful, and eye-catching character goods. Cute accessories are also added to the antennae of mobile phones, although mobile phones must be left in the changing rooms. The idea is that through these accessories women assimilate the value orientations of these goods and project that acquired value in their self-presentation. Similarly, high school girls wear mobile phones around their necks: "My phone is an extension of myself" (*Economist* 07/10/99). Women say that men at least can wear ties, shirts, and suits of their choice, although in practice there is not much variation in male dress (chapter 4). Women decorate their laptop computers with little "print club" stickers taken with their friends, and frequently change screensaver settings. The overwhelming choice at the time was the actor Leonardo DiCaprio after the huge success of the film *Titanic* in Japan. Multitalented Japanese *aidoru* who appear in soap operas, commercials, and game shows are also a popular choice (e.g., Kimutaku), as are cute animated characters.

Some older male managers who play around with computer related gadgets scan in family photos at home and bring it into work on a floppy disk to display as a screensaver. Women criticize these managers for being effeminate and doing something unbefitting of their age and gender. Women think of these acts of embodiment as feminine acts and a privilege pertaining to them. Little mascots and stationery displayed on women's desks are also an embodiment of their personal private space; they demarcate the zone between public and private—each subsumes the other and cannot exist independently.

Practices of Detection: Mēru Monitoring

Generally, most male managers are unaware as to how women spend their time in front of the computer; they seem to think all women are hard working. However, some managers *do* notice. Women complain of managers who pass behind them (deliberately to monitor behavior) as they write personal emails as an invasion of privacy.

Sakura-san is pleased to have partitions placed in between desks facing each other. These partitions were approximately 30 centimeters above desk height, about eye-level when sitting down. The manager sitting diagonally in front of her is very tall and unfortunately, he peers over the partition at her. This manager at times asks her to complete a task by talking over the partition, but often comes over to her desk to sit at the empty desk to her left. He approaches her with tasks at inopportune moments, while in mid-flow of composing a silly email to her close friends in the office. She quickly brings up a work document she has ready in another window for such occasions to replace the email document on the screen, but he has caught her out frequently. Whether she is cautioned depends on his mood.

This shows that management's powers of detection lack accuracy and systematic rigor; hence lack fairness in the assessment of workers. As email is a new form of technology in the office, management has yet to master it as a means of controlling and assessing workers.

Importantly, we now discuss how women's "tactics of evasion" (diverting time at work for work that is free and creative—de Certeau 1984) is turned against them by other female workers in an effort to exert judgment, surveillance, and power. Monitoring email use as an everyday technique of regulation is commonly practiced among female workers; rarely is it enforced by management on workers, unlike the everyday operation of hierarchy as we saw in the previous chapter.

Many women have Holmesean powers of detection. They have a heightened awareness of when other women write personal emails. Slackers are known as *sabori*. Women quickly utilized the frequency of emailing as a means to critique others. Complaints against particular women are raised in private conversations in the changing room and tea room. *Sabori* provides ample grounds for gossip. Knowledge of others is used to rank and critique individuals. However, as email records can be erased from the computer's memory, it is said, there is difficulty in retaining quantifiable evidence to point against the *sabori*. (Although, much to the worker's surprise, we find below that this is not the case.)

Tanabe-san can distinguish between two kinds of emails by typing speed: personal email compositions sounds faster, invigorated, and more labor intensive than work-related emails. She complains that managers who are oblivious to the extent of personal uses of email hold misinformed views of those women, who, in fact, do very little work all day. In some departments, some women have a work deadline at the end of the month, making it possible to do the bare minimum until then. Although somewhat misguided, emailing makes women out to be hardworking when all the while they are writing to friends. As Tanabe says, this is unfair and infuriating to women who do work hard.

As *sabori* is difficult to prove, yet more difficult to disprove, this surveillance mechanism (email monitoring) prevails among workers as a means to evaluate the performance of others. This ambiguity is precisely why detection is all the more effective. This practice produces an articulation of workplace values through gossip: mainly of appropriate attitude and conduct, and of rights and wrongs. Female workers evoke workplace values not only in connection to email but also in terms of appearance (chapter 4). This reinforces the management discourse. In this, the worker's condemnation of others can become outwardly hostile. By producing order, interestingly, this case shows how the power to control workers in this community is not only contained in the hierarchical ordering of the organization; power is diffuse and rules are implicit.

We can see how gossip in JCars functions in a wide sense. As Gluckman (1963) writes, variously, gossip allows for the competition for positions of power among leaders of small tribes (Radin 1927); maintenance of the unity of groups and their morality (Herskovits 1937; West 1945); and marking membership to an exclusive group (Coleson 1953). Other workplace ethnographies have noted how gossip functions to "resist" management through a "persistent gaze" turned upon the management by workers, or it is a form of "entertainment" of the female workers (Lo 1990; Ogasawara 1998: 84; cf., Scott 1985).

Those who gossip do so in order to put themselves forward as the better disciplined worker, by informing others who is right and who is wrong, thereby sustaining the values of the company and circulating shared knowledge.

Further, the purpose of email monitoring and ensuing gossip is a way of articulating feelings of envy. Individuals successful in *la perruque* (de Certeau 1984) are subject to gossip: those skilled in the art of disguise invite negative comments from the more conscientious workers who enact the management's view of the ideal worker under the strain of under-recognition. Hence, the discourse of management is appropriated in order to legitimize their critique. It is interesting to note how a medium that allows greater freedom of social interaction transforms into a space that is used by workers to reproduce the management discourse. There is logic to this. Expressions of envy are regarded as distasteful, as an egotistical emotion (Foster 1972), thus a "socially justifiable justification [is needed to permit confessions of] strong envy" (p. 165). If the emotion of envy is conflated with a discourse of efficiency and productivity as espoused by management, it seems more justifiable to both the accused and the accuser. At the same time, surveillance and gossip about email entails another form of competition that serves to legitimize and improve status vis-à-vis workers.

Co-Workers as "Other"

We explored how the company shapes workers' experience of the office in terms of a community through symbolic practices (chapter 6) and by representation of hierarchy in the open plan office (chapter 7). This is, we argued, because interrelationships are founded on dissensus and disorder (*midare*) rather than harmony (chapter 5).

Indeed, the construal of email use by workers as a private activity and the surveillance of email use that followed evoke a notion contrary to the sense of community imagined by managers, where, in fact, colleagues become "other" (*seken*) from the perspective of individual workers. The fact that fellow workers and managers become *seken* implies that they are members of the group with shared standards, but interrelationships are not perceived as being close, but neither are they complete strangers. *Seken* is highly sensitive to the shifting positions of individuals within social interaction and in *seken*, positioning is interchangeable and inconsistent. At times, workers in a section are *seken*, although in other contexts they become *nakama* (friends). We can see then that this construal of fellow workers as *seken* is nuanced and requires further explanation.

Seken[14]

The Japanese script for *seken* combines the two Chinese characters meaning "world" (pronounced as either *yo* or *se*) with "space-between" (pronounced as either *aida, ma, kan* or *ken*). The core features of the concept are as follows: *seken* refers to the appearance of the total network of social relations that surround an individual; it conveys the corresponding cultural norms and values that function to regulate social behavior and hints at how such relations and behavior are maintained. *Seken* is thought to be a concept native to Japan that has existed since the seventh century. It corresponds roughly to "*shakai*," the translated word for "society," derived from the West, which came into circulation in the Meiji period (1898–1920) as Western concepts, ideals, and values became popularized by politicians and intellectuals. "The public" is at times used as *seken*'s English equivalent. However, the two terms are by no means synonymous, a conceptual lacuna exists between "the public," with its universalistic connotations, and *seken* that, by comparison, when referring to one of its meanings—network, points rather more to a specific social context, or *aidagara*. Hamaguchi

Esyun (1985) discerns how interrelations that constitute *aidagara* include encounters that are functional as well as unintentional and nontransactional. Thus *seken* can be described as the sum of interrelations as a result of the accumulation of subnetworks of *aidagara*.

A diagrammatic depiction of *seken* clearly embodies the two core features given above. The model represents the interrelations between individuals as a stratified concentric structure: the individual in the center, the people known to the individual (friends, work colleagues, neighbors) in the adjacent ring, and people in society that the individual does not know (or strangers) in the outermost ring. *Seken* points to the body of people who fall in the mid-region between the two. This model also accounts for the *breadth* of relations that surrounds the individual in everyday life; this mirrors a macro level reality that extensive networks sustain many aspects of Japanese social and economic life.

Seken is a relational term with a spatial reference, and the relation between the self and *seken* is ambiguous and precarious because the boundaries of the term are flexible, relatively arbitrary, and dependent on context. This situationally determined feature of *seken* poses three practical implications for its use: first, there is no singular or set way of identifying who—friends or strangers—fits these positions at any one time. For example, a particular individual, say, x-san, might include another individual, y-san, among the *seken* category on one occasion, but depending on how x-san feels toward y-san the following day, y-san might no longer be considered *seken*. Second, the term *seken* does not necessarily correspond to a particular individual. It can also be applied, as in most cases, to refer to a group of individuals who are neither close nor other. In this way *seken* is highly sensitive to the shifting positions of individuals within social interaction, and in *seken*, therefore, positioning is interchangeable and inconsistent. Third, the substance of *seken* can differ, dependent on sex, age, social origin, occupation, level of education, region, and marital status. The *seken* referent is therefore constituted either by the relations between these properties, each of which has its specific value, or by a single pertinent one.

Seken is a relational concept that entails a comparison between self and social norms and ideals in the context of daily practice. *Seken*'s presence regulates the thought and hence behavior of individuals that brings them into alignment with society's standards. Japanese people on the whole take seriously the implications of deviating from *seken*'s standards and they continually and minutely adjust the inconsistency arising between oneself and *seken*. The term *seken* is in frequent daily usage in contemporary Japan, where it can be experienced by the individual as an omnipresent force, constantly serving to judge and regulate behavior in a collectivist society. This sense is conveyed well by anthropologist Takie Sugiyama Lebra: "In parallel to the 'face'-focused self, the *seken*-other is equipped with its own 'eyes,' 'ears,' and 'mouth,' watching, hearing and gossiping about the self. This body metaphor contributes to the sense of immediacy and inescapability of the *seken*'s presence" (Lebra 1992: 107).

As *seken* expresses a type of obligation and conformity to the group, it can be related to the dyadic concept of *tatemae* and *honne* that mean, respectively: rules that are natural or proper that have formed on the basis of group consensus, and, in spite of a display of conformity to the group the individual's true intentions (see Doi 1986). *Tatemae* is also an essential technique in the presentation of the self. It is acquired by individuals through the learning and judging of social codes and used to survive in society by reducing the potential for conflict. In this sense of technique, *tatemae* differs from *seken* that refers to either people, or to a controlling force. Furthermore, insofar as *seken* indicates social norms that induce the conformity of individuals, the concept can be related to the Western sociological and psychoanalytical notions of habitus, social fact/conscience collective, and the superego.

It is not entirely clear, even in practice, how the notion of *seken* operates: it is equally relevant to understand that the individual is somehow regulated by *seken* as much as the individual can regulate his/her own behavior in accordance to *seken*'s standards. *Seken*, insofar as it can be construed as a disciplining force, can be interpreted as functioning similarly to the sociological concept of habitus. The concept of habitus is a "structuring structure" that shapes the practice of people at the level of the unconscious through a process of implicit pedagogy (Bourdieu 1977; 1990). Habitus is an internalized concept. The operation of habitus and *seken* are similar in the way socially appropriate norms of conduct become internalized by individuals whereby their practice becomes shaped implicitly. Yet the difference between habitus and *seken* is that individuals are conscious of the presence and pressures of *seken* to a greater extent than Pierre Bourdieu claims about habitus; *seken* seems to have a greater force of control. Furthermore, the boundary of the term habitus seems more stable than seken.

The concept of *seken* is often defined as being unique to Japanese society, but similar accounts of forces of control that regulate the body exists cross-culturally. As discussed, it appears to fit descriptions of habitus, a French sociological concept, applied both to its provenance (Bourdieu 1984) and to Kabyle society (Bourdieu 1977; 1990). Marcel Mauss (1979), who originally described habitus as an encultured bodily way of behaving, indeed intended the concept to apply cross-culturally, and Mauss's examples of habitus were developed based around his observations of French and American society. Parallel examples of forces operating like *seken* are also found in ancient democratic Athenian society and in Victorian England. For example, Allen (2000) writes that the practice of naming and shaming, gossip, and the close scrutiny of others functioned in ancient Athens to keep people in their place. In the absence of any official punitive system possessing concrete techniques of control, order in Athenian society was produced by such discourses concerning punishment and the substance of the law. Discourse about order tended to have the desired ordering effect whereby such discourses functioned to perform an endless maintenance of distinctions, values, and meaning in society. In these examples, public opinion sanctions behavior.

Ultimately, *seken* refers to the relation between the individual and society. As *seken* regulates the behavior of individuals in relation to norms, an understanding of the way *seken* functions can be compared to the functioning of group norms and ideals outlined in Emile Durkheim's deterministic model of society, in his concept of social fact. To be precise, the relation between individual action and society is the object of theorization. For Durkheim, social facts, which consist of ways of thinking and behaving, are coercive forces that penetrate the individual without the individual perceiving that they do. The social fact becomes part of the individual's thought and behavior in a way that transforms the individual by somehow tying him/her to the group by providing norms and ideals. By believing in the externality of social facts, Durkheim treated norms as properties of collectivities. These norms functioned to constrain, and combined with his view that social facts are moral phenomena, he explained how adherence to moral ideals incites action. By connecting the three spheres of morals, norms, and action, Durkheimean sociology explained how sanctions and constraints regulate individuals' behavior.

At the level of an individual's psychic structure and processes, the psychoanalytic work of Sigmund Freud provides a comparable explanation of how *seken* regulates an individual's behavior. According to Freud's model, the repository of social norms within individuals, the superego, functions by holding the individual's desires within the bounds set by society. The superego includes two subsystems: an ego-ideal and conscience. The ego-ideal is the child's conception of what his parents will approve of; conscience is the child's conception of what his parents will condemn as morally bad. Both are assimilated by the child from examples and teachings provided by his parents. The ego-ideal is learned through rewards; conscience is learned through punishments. Freud's explanation of how these social restraints become internalized to form the superego contain a European, middle-class bias. However, it is possible to infer from this how the internal psychic process might work in adults when they are faced with social norms to which they should conform. In the case of *seken* the role of the parents in Freud's model would pass on to the social body as a whole.

The theoretical proximity between *seken* and Western sociological and psychological concepts—habitus, social fact, superego—would appear to illustrate the extent of commonality in the human condition. It is also clear that the concepts are by no means commensurate with each other due to inevitable cultural specificities arising from the regional scale of observation and conditions of analysis. Our understanding of *seken* and similar phenomena would profit from a body of future research that applies in-depth ethnographic methods to explore more comprehensively how *seken* works and impacts on individual daily lives in all spheres of social life.

Mēru and Seken

Email is beneficial for women at times, as Sakura-san says, as a medium to relieve work-related stress arising from office social relations, that is, unreasonable managers and generally disagreeable co-workers. In this context, women engaging in *la perruque* say that "when writing emails to friends, it's possible to do so without the self-consciousness of *seken*" (友達とメールを打ってる時は世間を気にしなくてもいい). She says "日本人は世間体を気にする、自分を他の人から見た視点," which she translates as "Japanese people are aware of the standard of decency considered from the public eye, it is how one appears from another person's perspective."

Clearly, email relieves the worker from the surveillance of *seken,* as email is a way of avoiding the surveying "eyes" and "ears" of *seken*, something that a telephone conversation cannot pull off. And yet, because email has this capacity to disguise the activities of the individual, this is precisely why email use is monitored by women in acts of surveillance against each other. Thus, as much as it relieves the worker from the prying eyes of *seken*, it makes workers aware of the omnipresence and pressures of it. It is understandable then that having this important source of escape put under surveillance annoys women.

The Disorderly Community

> People all too often make the mistake of thinking that the contents of an email, like things said over the phone, are private and not permanent. But it is not only possible for an employer to read all your emails and monitor every Internet website you visit, it is also perfectly legal…(*Guardian* 27/03/00)

After my fieldwork, the worker's practices of detection passed into the hands of management. This occurred by coincidence, following a server overload due to the prolific amount of emailing by the workers. At this opportunity the management conducted a detailed investigation into the cause. Although successful in their daily efforts to escape detection, workers were eventually caught out by the technology that allowed them to be creative in their practices. Emails leave traces. In this event management applied the work versus non-work classification to critique the behavior of workers. This generated a sense of anticommunity in JCars. Thus, eventually, email monitoring operated in two contexts: between workers, and management on workers. Let us read from Setouchi-san's email what was involved in this investigation:

> ...メールの検査はこの前突然行われてん、会社のメールサーバーがパンク状態で原因を調べたかったらしい...他の女の先輩（40人中10人位いる）の順位は分からんけど、女の人で3人位、10位以内やったで。　発表の仕方はグループそれぞれで私の所は順位を回覧されてん！！！超ショックやった。　その事に関して、ノーツ関係の仕事の人は反対して課長を注意したみたいやけど部長は結局は、メールは会社のものだから中を見ても関係ないって思ってるらしい。　発表された時はメールの回数も気を付けていたけど、　今は元に戻ったかもね　（笑）　でも、会社のでしないようにポケットボード（携帯できるやつ）か　携帯電話でコミュニケーションを取るようにしてる

[... The inspection of email happened unexpectedly the other day. Apparently they wanted to find the cause of the company's email server overload... I don't know the exact positions of my female *senpai* on this list (of 40 departmental members 10 are female). Out of the top 10 about 3 were women. The inspection results were announced within each group [section], and in mine they passed around a circular notice!!! I was so shocked. The team of developers working on Notes [Lotus Notes: an application] were opposed to the way this matter was handled and gave the *kachō* a warning, but apparently the *buchō* thinks that **Notes is company property and so they have a right to look at the contents irrespective of what the workers might think.** Around the time the announcement was made, I was careful with the amount I emailed, but now I think I've returned to my previous amount (laughter). But I [surreptitiously] carry a pocket board (the portable one) [version of palm pilot with built in email and Internet capability; like Blackberrys] and mobile phone at work so I don't have to use the company email.] (Setouchi 28/05/00; my emphasis)

Further, there was another incident in which the feeling of anticommunity was heightened. When the LAN system was launched, the IT department designed a space on the shared system called "worker's plaza" (*shain no hiroba*). (Note to reader: 場 [*ba*] means space. And "plaza" denotes a public space.) *Shain no hiroba* was intended to encourage sociability among members of the workplace community, to facilitate face-to-face conversation after initial contact via this bulletin board. Workers with similar interests could post electronic notices that would then generate face-to-face interaction. Also workers could use this space to advertise for-sale notices, and so on. Furthermore, this space was where members of the company could post messages and suggestions to upper management.

The factories have a similar bulletin board called 緑の広場 *midori no hiroba* (a green open space/plaza). The site evokes an image of an oasis as greenery is lacking in the factory (noted by a member of the personnel department). This bulletin board was designed to facilitate communication between workers and management regarding jobs and tasks, and airing of technical problems relating to the production process.

Thus, both bulletin boards encourage knowledge enhancement: social knowledge for *shain no hiroba*; skills-based knowledge for *midori no hiroba*. It will surprise the reader to know that *shain no hiroba* was closed down abruptly and unexpectedly. The workers were also surprised. This occurred in May 1998, the month I was in London during fieldwork. When I returned, perhaps because it had just happened, it was difficult to find out the details leading up to its closure. No one knew why the site had been shut down. Sakura-san said:

会社の文句とか書く人がいて問題になった。　人事部がつぶしたという噂がでたりしているけど そういう訳でもないらしい。　会社の中での力関係が問題になった。

[It became problematic because some people wrote complaints against the company. There's a rumor going around that the personnel department ordered it to be shut down but apparently that isn't the reason. The power dynamics in the company was the problem.]

The issue was terribly contentious and guarded by whomever I asked, hence I was forced to abandon my curiosity. However, post-fieldwork, removed from the site of research, through email contact, I obtained further details. The following is an extract from Sayama-san's email:

最近、その当時社員の広場を廃止した人に聞けたんだけどやっぱり表と裏があるみたい。表向きは...

　1) ノーツを社内に広めるって事で始めた社員の広場だったから その目標が達成されたから

　2) 社員の広場はやっぱり遊びみたいな感じだから仕事中のそのＤＢを見るのは駄目なのでは？って事らしいよ...

　でも、本当はね...広場にね、中国の人だと思うんだけど中国語で社員の広場に書いてたんだって。で、中国語でその内容に返事（意見）を書いた人がいたらしいの、その事を中国人に中国語で書かない方が良いって注意した人がいて、その人達が殴り合いの喧嘩にまでなったらしいねん（どう言い合いになったかはしらん）〔注意した人が公共の場だからみんなに解るように書くべきと言ったのか、中国の人は自由の場 だから いいじゃんと反論したのかそこまで聞けず〕

　それが人事部にまで行ったらしくて人事部から廃止するようにって感じで言われたか、システムが自主的に廃止したのかどっちからしいねん。　そうなったらしょうがない気がする。

[Recently, I was able to ask the person who abolished the workers' plaza. As expected, there seems to be a surface/false reason [*omote*] and a hidden/true [*ura*] reason. The surface reason is...

(1) The workers' plaza was started to familiarize workers with the use of Notes and as this aim was achieved they closed it down

(2) The workers' plaza has an informality about it, so the general feeling was that workers shouldn't be looking at that database during work hours, apparently...

But, the truth is...I think it was a Chinese person, who wrote messages in Chinese and someone wrote a reply (opinion) to that person in Chinese. Someone then warned these Chinese people that it would be better if they didn't write in Chinese and this led to a punch-up between them (I don't know what was said exactly).

{The argument could have gone like this: the person who issued the warning might have said that the **workers' plaza is a public space** and so they should write in a language that everyone can understand, and the reply might have been that the **worker's plaza is a free/open space** so people should be able to do what they like, but I wasn't able to ask for details.}.

Apparently, this issue reached the personnel department, and I don't know whether it was the personnel department who issued the closure or that the IT department took the initiative and closed it themselves. Given the state of things I think the closure was unavoidable.] (My emphasis)

As Sakura explains, between the management and workers, albeit unintentionally, a vast discrepancy in interests arose between the two factions, based on opposing interpretations of ownership and usership, and in the meaning of public space. The use made of this space threatened to undermine the authority and order created by the management. It opened up a space for workers to voice themselves, but the community could not cope with this exposure. Management certainly reacted quickly to implement control.

If the issues discussed on *shain no hiroba* had been limited to technical concerns, as it was in the factory's bulletin board, the management would have kept it running. However, as the discontents aired at the head office were personal in nature, and exposed racial or national tensions (Sino-Japan rivalry), management closed it because flexible systems open up flows of knowledge. This rupture to the usual order demarcates lines of exclusion, exposes the operation of power, and reiterates the implicit rules of practice in this community.

Although the site was set up to generate a sense of community, because the community is already fractured, it became an outlet for opinions that exposed the community for what it is: disorderly community. This does not resemble the ideal community espoused by management. Thus, the circulation of the wrong kind of knowledge was curtailed by the management. The row between employees was not left to settle in its own time; this was perhaps to prevent escalation of the dispute. I mentioned at the start of this book that there are always prohibitions on what one must not represent. Yet the plot twists and turns, as the management themselves contributed to the disintegration of their idealized community by dismantling a space where their ideals of sociality and open communication could be fostered.[15] This disjuncture at two levels—(1) among workers; and (2) between workers and management—seems rather unfortunate, all the more as common aims regarding sociality flourish among them. The management has always encouraged the transcendence of hierarchy in one sense, through deeper communication and understanding, yet the realization of this ideal encountered problems as in reality workers are circumscribed by strict channels of communication and ways of behaving within vertical relations (chapter 5). In effect, what these disruptions to the workplace community signal is not so much a dramatic shift in ideology as complications in processes of transition, as the operation of ideology or hierarchy moves from accustomed spaces of the office into new spatial dimensions of the Internet. At JCars the problems concerning email and bulletin board use arose one and two years after the technology was put in place in mid-1997, in May 1998 and May 2000 respectively. Invariably, all firms encounter teething problems ensuing from the installation of new technologies as firms adjust to the ways social interaction and work processes are altered by their use. In future, multilateral solutions might enter the management's repertoire as they strive to accommodate alternative procedures for resolving similar and related problems. We should remain hopeful that management would forge a new ethos by embracing new configurations in the culture of working life.

We might end dialectically: as much as the workplace community is shaped through discursive practices (Anderson 1991), what is exposed in this ceaseless and fascinating process is the impossibility of conflating a relation or social bond, this being-in-common, with common being or community (Nancy 1991: 29). We are alone with our subjectivity and together with others who are alone with theirs. But in our experience of the otherness of the other there is also a recognition of the otherness in ourselves; in relating with the other something changes within us, and this allows identity to be shared, thereby making community possible (pp. 33–4). Surely, a greater openness to embrace this understanding of difference is essential to the philosophy that shapes corporate ideology. In Japan the structural economic pressures at home and abroad in the 1990s led to the gradual breakdown of corporate-centered society in the recessionary period; Japanese society is said to be increasingly fragmented and atomistic; and like other countries globally, forms of work and economic life are diversifying, while new technology reshapes social relations and dynamics toward openness and networked structures. The Japanese workplace community will not cease to exist as we know it, but will continue to change shape ebbing with the flow of socio-cultural, economic, and legal changes.

Intermission

By now, all the direct rays of the sun had completely disappeared. The sky no longer held anything but pink and yellow colours: shrimp, salmon, flax, straw; and then this distinct richness could be felt to be fading away too. The heavenly landscape was re-created in a range of whites, blues and greens. The small corners of the horizon still enjoyed ephemeral and independent life. To the left, a hitherto unperceived veil suddenly came into view, like whimsical mixture of mysterious greens; these gradually changed into reds, at first intense, then dark, then purple, then dark purple, then coal-black...

Nothing is more mysterious than the series of always identical but unpredictable processes by which night follows day. Its mark appears suddenly in the sky, attended by uncertainty and anguish. No one can foresee the form that the resurgence of night will adopt on this unique occasion. (Levi-Strauss 1973: 67, my italics)

Such beauty. I leave the office and the social relations within it behind for another day. The chorus of *osaki ni shitsurei shimasu* (I'll be leaving before you) and *otsukare sama deshita* (well done, you must be tired) fading. The distances between stations are close together while the train is still in Osaka. The last stop in Osaka, before entering Nara Prefecture, where the train emerges from the underground level is Tsuruhashi. Here the opening of the doors in the evening invites the smell of *yaki-tori* (grilled chicken) and Korean style *yaki-niku* (grilled meat) into the carriage. As I am carried towards gentle suburbia, the workplace and the city seem distant but never so in my consciousness as I anticipate the next day at the office.

The seated passengers appear to be asleep, as they sway heavily, in stable unison, to the motion of the train as to an orchestral score. Most appear to be office workers. One elderly man I notice has a perfectly spherical cranium, the epidermis taut with the look of a rather charming, polished leather peach. I admire his cranium. The others are women who look to be in their twenties or thirties in suits, their make-up still intact, and salary men of various ages. The chatter in the crowded train during the early stage of the journey was taking place among co-workers as they spoke about the people in the office and various projects they worked on, before the conversation shifted to family life. Every evening there is one person who wakes up from this momentary slumber, just in time to race off the carriage at the correct stop. At the last stop on the line, where I debark, there are often men who are heavily asleep, who when shaken by the conductor do not stir. The life of the office worker (including social commitments), of whatever rank in the company or sector in the economy, is burdened with exhaustion.

Conclusions

Closing

At this point, I wish to express my gratitude to the generosity of the women and men at JCars who made this ethnography possible. In this book about their world, I examined the changes to women's status in workplaces and workers' awareness and needs that have been reshaped by the transition of socio-cultural values, which in turn impacted the way management maintained their ideal workplace community. My fieldwork was an opportune and unusual time to capture the structural shifts of the 1990s, in labor markets and employment practices in businesses, and in cultural changes in wider society affecting working life. The timing of my fieldwork from February 1998 to February 1999 allowed me to contribute new empirical data on the implementation of the revised 1997 Equal Employment Opportunity Law (EEOL), enforced April 1999; and on the social and work related consequences of installing IT infrastructure at JCars in the late summer of 1997. I also drew on other rare topics from which to theorize workplace social relations: symbolic and spatial practices. An unexpected topic that emerged during fieldwork has been the place of disabled people in white-collar offices, and the wider employment practices with respect to disability in Japanese businesses.

I was interested to draw out the complexities, contradictions, and concealments present in workplace knowledge about gender, status, ideology, and community, as expressed via language and embodiment. I drew on a wide range of contexts including the discursive construction of occupational role; the EEOL's impact on women's career paths; the idea of the "elite" worker and career mobility among men; "sister" relations with reference to other sets of vertical relationships within work groups; temporal and symbolic practices of gift-giving and conforming to company time; hierarchy and its spatial practices as well as workers' movements across different spaces.

My orientation to my topic is crucial to this investigation of gender, status, and ideology, and these form the unique features of this book: First, when analyzing gendered aspects of work the natural tendency is to explore relations occurring between men and women (e.g., Ogasawara 1998) to the exclusion of the most frequently occurring pattern of interaction, and by this perhaps the most significant in terms of its importance to the workers' everyday lives, that between same-sex members. It was evident, that, in order to convey depth of experience about culture of work, it was necessary to view the workplace not so much split down gender lines of men versus women, but as more subtle, interrelating, and complex, in ways that more accurately resemble working relationships. Thus, it was vital to explore women interrelating with men *and* with other women. Analyzing female interrelationships in their own right (not in terms of whether they can form organized protest) can add to our knowledge of organizations and workplace relationships, particularly in this time when the nature of work and occupations, and societal attitudes to working women are changing.

Second, building on the need to recognize coexistence of dissonance and harmony that is often missing in the discussion of workplace community in Japan (Turner 1995), I began from the premise that relations within the community are predisposed to disorder (*midare*) and differentiation, which mirrors wider societal change, and from this state management and workers effectively shape order and their ideal community. A further point in relation to coexistence (partially made above in connection to the importance of analyzing same-sex interaction) concerns linking theory with my subject's practices, which involves the resolution of the conventional oppositions that structure culturally dominant representations with regard to gender, identity, status, and work (Allison 1995; Bachnik 1998). These are men versus women; career track versus clerical track workers; inside versus outside; management versus employees; work versus play and other non-work activity; and, formal versus informal spheres of activity. My task was to show how knowledge is produced through the dialectical tension between coexisting categories via a discursive approach recognizing multiple voices (Kondo 1990a) and the place from which the subject speaks. In addition to bringing Practice Theory into conversation with anthropology of Japan and the sociology of work, I wished to reinterpret the notion of hierarchy within anthropological analyses of the Japanese organization. The ethnographic and theoretical points outlined in this book are summarized below.

The examination of recruitment methods and the tracking system at JCars showed that the impact of restructuring was felt most strongly by women compared to the men of their cohort in the form of continuing differential treatment endured by women in recruitment and promotion. This was explained in connection to wider socio-economic changes such as the increasing pattern of restructuring by layoffs and voluntary early retirement taken in large Japanese companies, the increased acceptability of occupational mobility among male workers in their twenties and thirties, and the introduction of internship schemes for graduates. Further, the revised 1997 EEOL failed to deliver equality as wage and promotion differentials were legitimized by the management through arguments of cost-efficiency in this recessionary period. The numerous constraints experienced by women discouraged them from attempting to reverse their experience of gender inequality by improving their status in official, objective terms through promotion to career track positions. These constraints were not only policy related, but included other women's hostility toward career-mindedness in women, and the importance of social, cultural, historical, and professional norms associated with the role of women as wives and mothers. However, these norms supporting corporate-centered society also affected men, who worked long hours while experiencing social pressure to be breadwinners. Thus, the equal sharing of responsibility between men and women in the home was difficult to achieve, and, consequently, most women took early retirement upon marriage. But the trend for early retirement is altering as a third of women employed at JCars have had careers of 10 years or longer. Therefore, women's career aspirations are growing. Yet, in structural terms, most women's status appeared to be low.

Most crucially, however, at the same time, the occupational role of female clerical track workers was undergoing a significant change. This change was a by-product of the restructuring of personnel through the recruitment of fewer graduates (particularly female graduates) and by the increased levels of automation in the office. Female clerical track workers were thus given equivalent work to that of the career track workers. This experience of change within occupational role was directly linked to perceptions of status. I gave a detailed analysis of women's *experience* and their interpretation of status positions and showed that this focus on experience posed a more egalitarian picture of status differences than status depicted through purely structural means. Identified, too, in this analysis of experience, was that standards and expectations of gender equality vary considerably among women (some women were content with evaluation and appraisal whereas others sought remuneration by wage increases and promotion). I made the point that if experience shows that standards and expectations of gender equality vary among women, yet, if this shift in occupational role identifies a new process of inclusion at work in late modernity, even if the meaning of this experience is debatable and contested among women, we might still view this inclusion as a step forward for workplace gender relations.

Furthermore, it was clear that managers who did not occupy prominent positions within the management hierarchy influenced women's views of status. The former did so by allocating work involving greater complexity to women in the clerical track, and by encouraging some clerical track workers to apply for career track positions. Although this change in occupational role was not reflected in most women's official ranks, it changed the way women conceptualized and experienced their status (as being equal to career track women or men of their cohort). Therefore, the "dominant class" can be misconstrued as being unified yet, clearly, it is comprised of conflicting interests or "cultural practices" (Bourdieu 1984). And where it has been carefully pointed out in workplace ethnographies that workers should not be viewed as a homogeneous group (Matsunaga 2000; Turner 1995), this book has advanced a further point that heterogeneity among the management ought to be recognized as well. Moreover, not all managers thought of women as being outsiders in the office, and women themselves certainly did not consider themselves as outsiders or of low status all of the time, thus, in the discussion of women's status the removal of traditional dichotomies between worker/manager, women/men, and inside/outside begins to make sense.

Importantly, we saw how status is always in the process of being made and remade. Simply put, how workers defined their own status, as well as that of others, exemplified a struggle between incorporating the dominant system of meanings into one's own way of interpreting status and attempting to break free from becoming subsumed by dominant values (Bakhtin 1981). Status negotiation in various contexts of social interaction was critical to office life. I examined the way men's careers were talked about among women and among men in relation to wider discourses of masculinity that touched on notions of class and status of "elite." Further, I explored alternative means of assessing status among women and men such as appearance, dress, and popularity. As appearance was significant to the management discourse it also became a means through which women competed for status among themselves. This connection between status and embodiment of ideals was also explored in connection to acts of surveillance workers turned on other workers. In other words, I showed that surveillance was not a privileged mechanism used by the dominant group (de Certeau 1984; Foucault 1977). Surveillance among workers took the form of monitoring and criticism concerning the frequency of sending personal emails. This exploration of use of email and proprietary Intranet contributes to both ethnographies of Japanese workplaces and cross-cultural studies of the Internet.

The examination of emailing practices highlighted the way workers construed their relationships to fellow colleagues as *seken*. This fits with the experience of the workplace community as disorderly and differentiated. *Seken* was a relational term with a spatial reference, where the relation between the self and *seken* was ambiguous, precisely because the boundaries of the term were flexible, relatively arbitrary, and dependent on context. I suggested that *seken* is comparable to the notion of *habitus* (Bourdieu 1990, 1997; Mauss 1979) insofar as both could be construed as a disciplining force. *Seken* is not uniquely Japanese but a cross-cultural phenomenon.

In the discussion of spatial practices, I examined the part hierarchy itself played in maintaining the firm-as-community ideology. These were to promote efficiency by exposing the pace and content of information flow; to enable the surveillance of workers; and to enable a display of management authority relative to workers through the infrequent but spontaneous and effective staging of public spectacles. Furthermore, it was shown that hierarchy is not a rigid concept; it is open to use in any context, by management and also by workers, as a way for individuals to reaffirm their own status and rights vis-à-vis others. The workers' spatial practices was discussed in relation to gendered spaces of the tea room, changing room, and the smokers' room. This exposed the processes of contestation and reaffirmation of status evident from the way senior women criticized junior women for occupying the tea room and changing rooms at irregular times throughout the day. Moreover, among the junior women, new practices such as smoking within the office premises during the lunch hour, and the words (as action) of Muro-*kachō*, the manager who prevented me from carrying out an outdated gendered task of wiping the desks in my section, implied that the space of the office was shifting both conceptually and literally, toward becoming a space for men and women. This analysis of spatial practices enabled us to see the manifold ways in which equality within gendered and hierarchical relations are negotiated.

I was also concerned to show the alternatives to hierarchy as a trope used to maintain the community. I demonstrated that the hierarchical ordering of social relations *per se* does not foster a sense of community, rather it was via management pursuing an idea of community imagined in a particular style that this was achieved: through practices imbued with symbolic significance that engaged all members of the workplace (management and workers), the management was able to inculcate workers with their ideals and values of commitment to the workplace.

The starting point for this argument was, in contrast to traditional representations of the *nihonjinron* school and conflict and resistance models of workplaces, to see the company in terms of a symbolic community where interrelations between workers were characterized by dissensus rather than harmony. The accounts of workers exemplified the extent to which underlying assumptions about the nature of social relationships and the requirement to demonstrate appropriate conduct within them was contested and questioned (particularly among junior workers). Cohesion between members within this symbolic community existed only as an ideal, which required a performance in an enactment of ideal behavior to demonstrate the knowledge of social structure within vertical and horizontal relationships. Thus, it was argued that hierarchy does not naturally create harmony and consensus because workers conscientiously manufacture and maintain idealistic social relations. Thus, alternative ways through which the management discourse constructed an image of community and enforced their ideals via symbolic means were examined.

The regulation of the individual's notion of time through an imposition of the management's official spatialized and objective time was essential to this symbolic community. At JCars, time was loaded with the management's values. The duration of time spent in employment at the company was conceptualized as a time of fulfillment and enrichment, and the willingness to work overtime was indicative of a worker's dedication and was used as a way of assessing levels of commitment and awarding merit. What the company's notion of time produced was "bodily dispositions" (Bourdieu 1990: 75) for the workers who acted in accordance with it. Therefore, time was used to nurture a sense of community: time carried and conveyed emotional connotations, enabling managers to solicit obligation and loyalty from workers, while at the same time inculcating workers with their ideals and values of commitment to the workplace. Through three examples of male workers' notions of time, I showed that the degree to which individuals conform to the management's values depended on an individual worker's age, interests, the work ethic upheld in their department, the importance of work in relation to his personal life, (dis)ability, and career plans.

It was shown that gift-giving and the organization of birthday lunches led to the expression of emotions appropriate to fostering a sense of community. The exchange of perishable food items as gifts to colleagues in the department generated a feeling of obligation at a personal level and acted as a form of public sanction. Thereby, appropriate behavior was demonstrated and helped to maintain a sense of community. These ritualistic events were important to the workplace, as social relations were to an extent superficial, built on a slippery premise of membership to the group. Thus, the offering of gifts acted to compensate for the fragile nature of these bonds. In other words, gift-giving did not function to strengthen emotional ties, rather, it functioned to maintain the form of community.

However, the practice of seasonal gift-giving from juniors to seniors focused on expressions of gratitude within emotionally tied relationships, the aim of which was to create stronger bonds. Seniors taking juniors out to lunch or dinner and taking on the cost (*ogori*) was a way to mend vertical relations that become ruptured in the course of working together, presenting an occasion to restore the senior's rank, status, power, and prestige (in his own eyes). At the same time, there is no doubt as to *ogori*'s material effect as relations in the vertical relationship became soothed. *Ogori* thus has a double function at individual and group levels. It was shown that the temporary nature of this restorative effect requires its repetition.

In short, the necessity of repetition within these ritual-like activities such as conforming to time and gift-giving highlighted the fragility of the management discourse and the extent to which the cultural mores of the office needed to be made concrete. This goes for interpersonal feelings in vertical relations as well. By discussing the role of the management discourse in the maintenance of a symbolic community, I showed first, how ideology is ingrained within structures of practice, and without the repetitive practice, the ideology is redundant (Bourdieu 1977; 1991), and second, that flexibility existed in the way individuals might interpret the meanings of seemingly concrete ideologies.

Two reasons account for why I have referred to Nakane's (1970) concept of hierarchy, developed by looking at the function of enduring sets of vertical ties between workers within large white-collar workplace organizations. First, the similarity between Nakane's ideas and the management's expectations of conduct within social relations at JCars. Second, because, in one way or other, most Japanese workplace ethnographies seem rooted in Nakane's model. Nakane's vertical model is one variant of the "group model" (Befu 1980), which is a popular conceptual tool of structural-functionalist analysis in the anthropology of Japan. Nakane's model explained the stable, consensus-bound character of society, maintained through patterns of vertical sets of relationships. The force of Nakane's argument sits comfortably within explanations of the total structure of the group, based on patrilineal forms of social organization of the traditional Japanese household, surviving despite the rapid, postwar, socio-economic modernization of Japan.

Thus, in parallel to the management's notion of hierarchy, the foundation of this group model that I questioned was its supposed "rigidity" (Nakane 1970), where change is contingent on a "drastic event" involving the reordering of workers in the hierarchy or of a total "disintegration" of the group. The well-noted criticism of Nakane's model concerns the fact that change cannot be accounted for, mainly as the unique structuring feature—hierarchy—synonymous with the group itself—would be eradicated. Effectively, I have explored how hierarchy is not rigid. Rather, it is a fluid concept that can be variously remade and reinforced in different contexts of interaction by management and through the practice of workers. For Nakane, the problem is bound up in her construal of structure: by assuming hierarchy as an entity on its own, she attributed a mechanical function to hierarchy in its maintenance of stability and consensus of the group/society, when, in fact, hierarchy was actively constituted by and through social action (Bourdieu 1977, 1990).

Yet, undeniably, the hierarchical structure was a valid element contributing to a worker's experience of life at the company; equally, I showed the extent to which it was contested in discourses of status negotiation. So why does hierarchy present such an analytical dilemma for the anthropology of Japan? Presumably, because hierarchy serves an analytical purpose; however, it lacks the capacity to capture social life as it is actually practiced (Asad 1979). Rules and models cannot determine a society: its structures and how it functions. At best, they can only be applied to an understanding of society. In order to say that models are adequate representations of social life, social practice needs to be perceived rigidly to coincide with its tool: as capable of exemplifying sets of rules or as abstractions from a system of forms. Yet, social practice and experience are different: actual practice and experience cannot be predetermined to the extent made possible in rules posed by models.

Ludwig Wittgenstein (1997: §185–90) would argue that even rules do not determine outcomes. With respect to rule following: Wittgenstein's main point is that no rules can be induced from observing a finite number of examples. For instance, it is impossible, say, to induce what rule somebody is following when you watch them count up from one. Even if you watch them count up from 1 to 99 you cannot guarantee that the next number they count is 100, as there are an infinite number of counting rules that give you 1 to 99 initially, but, something other than 100 the next. Thus, it follows, that, despite hierarchy being a fact, any model, for example, that attempts to explain the rules of hierarchy in ordering social practice is philosophically incoherent as no set of empirical observations (even those of anthropologists) can determine a social rule. People do not follow a rule governing the practice of hierarchical structure all of the time, because social action is a creative process (de Certeau 1984), which like language is given meaning through "use" (Wittgenstein 1997).

This brings my argument back to Asad's (1979) point that representations of social life made by the anthropologist laid out neatly in social models differs from the perception the informant has of his/her everyday experience of social life. This point can be illustrated by the difference between (visual) experience and a picture (read model) of it:

> ...[I]f water boils in a pot, steam comes out of the pot and also pictured steam comes out of the pictured pot. But what if one insisted on saying that there must also be something boiling in the picture of the pot? (Wittgenstein 1997: §297)

Nakane (1970) might say, this steam or the pot, being a distinctive feature of the picture, is equivalent to hierarchy: a feature of social life objectified in her model. Although, it is not possible to see the boiling water in this picture, it is a necessary condition for seeing the steam rise out of the pot, and to go back to Nakane's approach to culture, the link between the boiling water and the steam is not made. The boiling water, refigured as the essence of social life, poses little significance to Nakane because, although it can be experienced and its presence imagined, it cannot be pictured in her picture/model of Japanese society.

In summary, the workplace community was shaped through discursive practices. The bittersweet irony of maintaining ideals in the face of a disorderly community was captured when relations were unmasked as how they actually are, in two events that forced management to intervene in its community. First, the management's handling of the worker's overuse of email that led to the malfunctioning of the corporate email server. Second, the management's response to the "misuse" of the bulletin board on the corporate Intranet. Both underscore the challenges faced by management when channels of communication are opened up by technology use. In case of the latter, the bulletin board was introduced with the aims of increasing sociality between workers and to improve communication between workers and management. However, as events showed, the introduction of IT, in addition to literally providing an alternative space for interaction, opened up a space where the meaning of public space and freedom of conduct within this space was fiercely contested. The kind of knowledge that became exchanged was deemed inappropriate by management. This led to the closure of this space. Unfortunately, for all concerned, it was a space where management's ideals of sociality and open communication could have been fostered: the management has always encouraged deeper communication and understanding between workers. Thus, these developments show the complex process of transition as the operation of ideology/hierarchy moves from accustomed spaces of the office to new spatial dimensions of the Internet. All firms encounter teething problems in relation to new technologies as firms adjust to the ways social interaction and work processes become altered by their use. In future, multilateral solutions might enter the management's repertoire as they strive to accommodate alternative procedures for resolving similar and related problems. We should remain hopeful that management would forge a new ethos by embracing new configurations in the culture of working life.

Impossibility of Closure

As this ethnography prepares to close, I am prompted to contemplate (already reminiscing) the relationship between the ethnography, ethnographer, and reader. I open *The Book of Sand* (1975) and go to "The Other." In this parable—masterfully examining the themes of the past, memory, forgetting—Jorge Luis Borges encounters himself as a younger man along the Charles River on a bench that existed in two times and two places: "The encounter was real, but the other man spoke to me in a dream, which is why he could forget me; I spoke to him while I was awake, and so I am still tormented by the memory." Like the elder of this parable, the ethnography narrates in the present a past; like the younger, my subject's past enters the present, while life continues to unfold; together with the two men, the reader of the monograph is at the enigmatic interface of past and present, disrupted in time. Are we not all in the place of dreams and memory?

Ethnographer to the ethnography: a relationship of a specific past that is still very much in the present from which we can read future trajectories:

> Time is, therefore, not a real process, not a real succession that I am content to record. It arises from my relation to things. Within things themselves, the future and the past are in a kind of eternal state of pre-existence and survival.... What is past or future for me is present in the world. (Merleau-Ponty 2005: 478)

And so it was.

After leaving the field, I kept in touch with a handful of women via email and letters. I saw one or two periodically, at the end of the year, a time when I visit my family in Japan. In October 2003, I was in Tokyo and Osaka for a consultancy project; it was then that I revisited JCars. It felt as though I had never left. All remained incredibly familiar. In the overseas department, mostly everyone who was there during my fieldwork was still there. Men transfer much more and I didn't recognize some of them, but all the women were the same, bar one or two. The women said variously: "you look more grown-up," "when you were an intern your hair was so short," "you've turned into a beautiful woman."

I interviewed two managers in the personnel department who were there during my fieldwork: Yuasa and Tomiyama (-*shunin*: I was unable to verify change in rank lest I were to appear rude). After my fieldwork, JCars went through some restructuring. To create a flat organization they cut back on the number of sections and departments. Managers are feeling more pressure as a result. To cut costs some departments relocated to a site in a neighboring Prefecture. From 2004, they will try out a new merit based promotion system. Although, in a questionnaire survey, only 50 percent of employees voted in favor as they feel that "merit" is assessed arbitrarily, nonetheless they will press ahead. 17 women out of 300 permanent employees are on the career track. The Office of Women and Youth Employment wish to abolish the tracking system, but the personnel department wishes to retain it. They insist that women are better at repetitive tasks such as data entry. The personnel department conducted a survey on gender discrimination within JCars. The result showed that there was no discrimination against women. The female career track workers feel they are able to use their ability to full effect. Sakamoto-*shunin* has been promoted to *kachō*, managerial class. Her tenure is coming up to 21 years, and Ashimoto-*shunin*'s tenure is 18 years. Moreover, marriage is no longer the ultimate goal for women. Women do not retire on marriage; on taking maternity leave they return to work after one year. Alternatively, they work part-time in other subsidiaries owned by JCars. It is rare to leave the workplace to become a housewife, although some take that route. Furthermore, non-regular workers, agency hired temporary employees, have become reputable for having better business manners and computer skills than new graduates. Therefore, JCars prefers to recruit from among agency staff.

I also caught up with my contact, a female employee, in the personnel department at a multinational electronics company headquartered in Osaka (Electron). Whereas JCars flattened their structure by reducing the number of departments, Electron readdressed hierarchical cultural norms by mounting a campaign to change the way workers are named in association with their rank. Instituted by the personnel department, status indicators are removed from the suffix in exchange for a generic "*-san*." This is Electron's attempt to encourage frank communication by ironing out differences in the perception of others by age, status, and experience. It is enforced through monitoring and the payment of a small fine (100 Yen) every time "*-san*" is forgotten. My contact confirmed that women's tenure at Electron is displaying the same trends as JCars. Women don't leave because of marriage; rather they leave to study or work in other companies. Childcare is subsidized by the company, and nursery fees are part reimbursed. Women can also reduce their hours for a while (*tanshuku kinmu*). My contact, a returnee, a graduate of a prestigious Japanese university, at the time in her early thirties, said marriage is not a realistic situation for her, she prefers to work.

At JCars, Mori-san, who, as the reader will recall, transferred to another section in the overseas department, has been in her subsequent job for five years. Her work is the same level in terms of responsibility and difficulty as a section head. Her job deals with international trading law. Now she desires change. She feels she has achieved her goals in this position. Interpersonal relationships are not going well anymore. She spent three years preparing for the GMAT examinations in order to apply to study for a MBA in the United States. At the crammer school, she met people of the same age who are sponsored by their companies to obtain these qualifications. She widened her aspirations by mixing with them. However, she lost interest in pursuing the MBA. Subsequently, over a six month period, she applied for jobs in external companies (some with links to JCars) and a few offered her positions. This was reassuring on two counts: First, the external (read objective) evaluation gave her confidence to know that her achievements are valuable. Second, the job market for regular employment is flexible; as a woman it is possible to change jobs after 30 years of age (she is 31). In a few months Mori is leaving JCars to join her new company. When she was 29 and 30 she faced strong family pressure to get married, but that passed. Age 30 is a turning point—*jinsei no kawarime*. She no longer speaks or goes to lunch with her *dōki* group. Half are still at JCars.

Hori-san always said that she wanted to leave JCars. She aspired to a varied career and envied the *freeter* work style. She talked about going abroad and marrying a foreigner; he would take her away from Japan; although in her year abroad during junior college to study English she did not adjust well to life in the United States. Hori left JCars and worked as an assistant at a language school in another English-speaking country for 10 months. She used her knowledge of working in the personnel department. She returned to Japan to work for another company, a consultancy, something much closer to her trendy nature.

Ando-san who has a strong interest in development issues won a government scholarship in open competition. She took two years' leave from JCars to work on a project in South-East Asia.

Araki-san goes abroad once a year to dance school. She teaches classes in Japan. The job at JCars is a means to get by.

Miura-san still lives in the company dormitory, and has two years left until he must leave (i.e., marry). He spends a great deal of time and energy supporting the labor union. He is a rising star, an important figure in the accounts department. It is rumored that he will be the first in his year group to be promoted. One member of Miura's *dōki* group from the personnel department was selected to go abroad to the United States as a generalist trainee manager.

Kubo-san is now a member of the special projects development team in the IT department. He remains without title, although people of his age have been promoted to *shunin* rank. He is developing software for the personnel department: if this is successful, under the new assessment system, he is likely to gain recognition for his work and receive promotion.

In 2006, JCars merged with another company (a multibillion organization). There were no staff layoffs. My father was on the management team of the joint venture that led to the merger. Around that time my father's *senpai* asked my father to join him at another company. With that new challenge my father retired from JCars; he is now *senmu-torishimari-yaku* at his new company. His work continues to take him across Europe, America, and Asia.

At the end of fieldwork my father kindly said that it must have been a difficult year for me, but he also gained valuable knowledge from the experience. While he was trying to gain permission for me to begin research at JCars, he spoke out for his beliefs. When his superiors responded favorably, it encouraged him to be more outspoken in future.

At the end of my visit to JCars, I caught up with my father's best friend from his *dōki* group, now the head of the overseas department. His words:

> You should leave London and come to live in Japan...Your father misses you, but doesn't say such things directly...Do you know how much he looks forward to your visits? He really worries about you, although he's unable to tell you that himself...You should give him some grandchildren because his job takes so much out of him...Your father needs *iyashi* (healing)...

The conspicuous lack of language between father-daughter.

Equally moving was a letter that arrived at my address in Maida Vale. My friend at JCars, wrote:

私らさまよってる子羊やな
オアシスからその次の架空のオアシスへ

[We are wandering lambs, aren't we
From one oasis to the next imaginary oasis[1]]

Oasis = exotic spaces: jobs; social relationships; a geography; a feeling; imagination. Oasis for the foreigner[2] = home—a place of *me*, who belongs solely for *me*: no longer the constant Other, the *female* Other—relentlessly beautiful, mysterious, and unknowable; feminine, fragile, and vulnerable; dangerous yet diminutive: neutral—only to face an apolitical indifference: stasis and permanence—take root, a place/partner once lost, now ready to be regained: fullness without traces of melancholy: that affiliation—to embrace a *we*, no longer solitary.

Oasis = is it within us?

Appendix

Appendix 1 Brief Timeline of Multisited Fieldwork

Dates	Places
Mid-January 1998	Arrive in Nara City, in the Kansai region of western Japan
Early February	Begin fieldwork in the personnel department of JCars at the head office in Osaka City
March	I also begin going to other companies in central Osaka City to conduct interviews with returnees
Early April	Attend a week-long induction seminar for new recruits at JCars' conference facility near one of the manufacturing plants in Nara Prefecture
End of April	Go to the international head quarters at the head office of national newspaper in Tokyo for interviewing and participant observation
Early May	Return to London to gain reflective distance from the field
End of May	On my return, I am given a seat in the overseas administration department, while retaining my seat in the personnel department
Early February 1999	Time to pack up the anthropologist's trunk and return to London
	☆ Includes spaces of email, intranet, and bulletin board

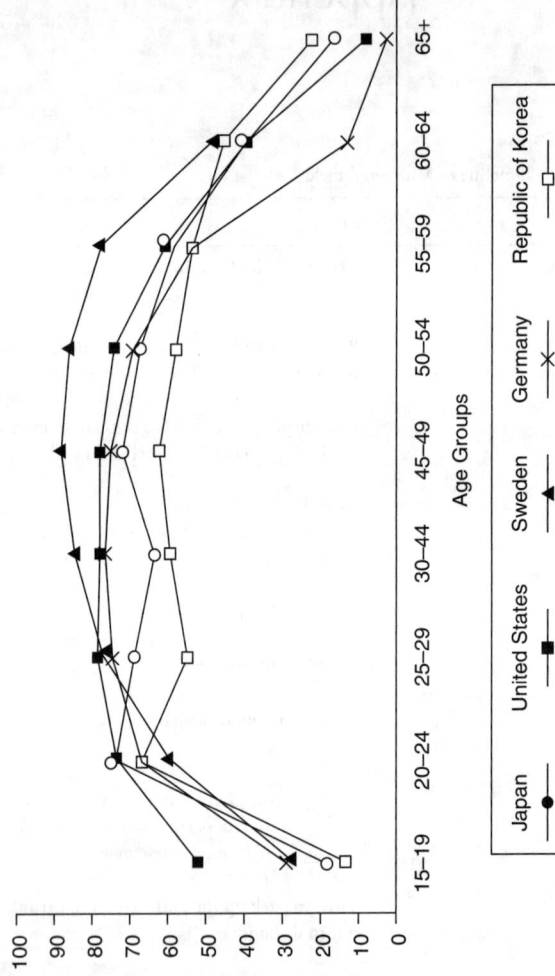

Appendix 2 A Graph to show the Labor Force Participation Rate of Women by Country and Age Group, 1997

Source: Compiled from "Present Status of Gender Equality and Measures," Sōrifu, 1999.

対象　　24歳　　　　　女性　　あなたの誕生星により定められし生涯

基本性格: 長所 *** 従順、反省的、情感が豊かで犠牲心があり、融和的です
　　　　　短所 *** 心配性で決断力に乏しく、意志がそれ程強くありません
　　　　　補足 *** マイペースを守る・・・周囲の人間との関係を円満に保つために自分の気持ちを犠牲にしようとは思わない

深層意識: あなたは　　　風　　の流れに属し「　野心　」こそがあなたの原点です。
　　　　　あなたの行動は、全てこれを基準に動いています。

愛情:　　あなたの愛情の星は　光っています

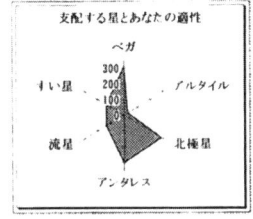

支配する星とあなたの適性

ベガ・・・・・・・・出世力
アルタイル・・・・・・幸運度
北極星・・・・・・・・金運度
アンタレス・・・・・・芸術性
流星・・・・・・・・・学術力
すい星・・・・・・・・浮気度

★運にたよらず実力でいきましょう

★あなたの生涯★

1999年	25歳	健やかに人生を送っています
2000年	26歳	健やかに人生を送っています
2001年	27歳	健やかに人生を送っています
2002年	28歳	おめでとうございます！あなたは結婚しました
2003年	29歳	健やかに人生を送っています
2004年	30歳	健やかに人生を送っています
2005年	31歳	健やかに人生を送っています
2006年	32歳	健やかに人生を送っています
2007年	33歳	健やかに人生を送っています
2008年	34歳	健やかに人生を送っています
2009年	35歳	健やかに人生を送っています
2010年	36歳	健やかに人生を送っています
2011年	37歳	健やかに人生を送っています
2012年	38歳	健やかに人生を送っています
2013年	39歳	健やかに人生を送っています
2014年	40歳	健やかに人生を送っています
2015年	41歳	健やかに人生を送っています
2016年	42歳	健やかに人生を送っています
2017年	43歳	健やかに人生を送っています
2018年	44歳	健やかに人生を送っています
2019年	45歳	健やかに人生を送っています
2020年	46歳	健やかに人生を送っています
2021年	47歳	大きな病気にかかります。
2022年	48歳	健やかに人生を送っています
2023年	49歳	健やかに人生を送っています
2024年	50歳	健やかに人生を送っています
2025年	51歳	健やかに人生を送っています
2026年	52歳	健やかに人生を送っています
2027年	53歳	健やかに人生を送っています
2028年	54歳	健やかに人生を送っています
2029年	55歳	健やかに人生を送っています
2030年	56歳	健やかに人生を送っています
2031年	57歳	健やかに人生を送っています
2032年	58歳	健やかに人生を送っています
2033年	59歳	健やかに人生を送っています
2034年	60歳	健やかに人生を送っています
2035年	61歳	健やかに人生を送っています
2036年	62歳	健やかに人生を送っています
2037年	63歳	健やかに人生を送っています
2038年	64歳	健やかに人生を送っています
2039年	65歳	健やかに人生を送っています
2040年	66歳	健やかに人生を送っています
2041年	67歳	健やかに人生を送っています
2042年	68歳	健やかに人生を送っています
2043年	69歳	健やかに人生を送っています
2044年	70歳	健やかに人生を送っています
2045年	71歳	健やかに人生を送っています
2046年	72歳	健やかに人生を送っています
2047年	73歳	健やかに人生を送っています
2048年	74歳	健やかに人生を送っています
2049年	75歳	健やかに人生を送っています
2050年	76歳	健やかに人生を送っています
2051年	77歳	健やかに人生を送っています
2052年	78歳	健やかに人生を送っています
2053年	79歳	健やかに人生を送っています
2054年	80歳	健やかに人生を送っています
2055年	81歳	健やかに人生を送っています
2056年	82歳	健やかに人生を送っています
2057年	83歳	健やかに人生を送っています
2058年	84歳	健やかに人生を送っています
2059年	85歳	健やかに人生を送っています
2060年	86歳	健やかに人生を送っています
2061年	87歳	健やかに人生を送っています
2062年	88歳	健やかに人生を送っています
2063年	89歳	大切な者を失います。
2064年	90歳	健やかに人生を送っています
2065年	91歳	健やかに人生を送っています
2066年	92歳	健やかに人生を送っています
2067年	93歳	健やかに人生を送っています

平凡な人生でした。しかし、それが一番幸せだったんだろうと思いつつ・・・
2068年　　8月　　　　20日　死亡

Appendix 3　　Uranai, Fortune Telling Based on One's Life Course

Glossary

Aidagara (間柄). Interrelations in a specific social context, includes encounters that are functional, unintentional, and non-transactional

Ba (場). A place; a space; a location; and in the context of social relations the term indicates one's situational position

Basho (場所). A place

Buchō (部長). General manager: usually in their forties

Chiiki gata (地域型). Regional type: Following the diversification of employment practices, a new type of career track option available for women in the financial sector (2006). The position involves transfers to any of the company's domestic subsidiaries only after the woman has reached manager level, which is unlike male workers who can be asked to transfer at any point in their career.

Chiteki-shōgaisha-koyō-sokushin-jirei (知的障害者雇用促進事例). An employment law targeting white-collar workplaces regarding the recruitment of workers with disabilities.

Chōtei-seido (調停制度). Arbitration policy

Chūto-nyūsha (中途入社). Mid-term recruitment; the new recruit who joins the company at times other than April when the majority of new hires begin work

Danjo-koyō-kikai-kintō-hō (男女雇用機会均等法). Equal Employment Opportunity Law

Dankai-sedai (団塊世代). Postwar, first generation baby boomers born between 1945 and 1952

Dankai-junia-sedai (団塊ジュニア世代). Designates those born between 1973 and 1980

Dōki (同期). Group comprised of workers of the same year of entry

Dōryō (同僚). Individuals of the same rank or horizontal stratum

Doryoku-gimu-kitei (努力義務規定). Regulation enforced by exhortation

Enko (縁故). Personal connections, cf., *kone*

Freeter (フリーター). [1] Underemployed or freelance workers who are not in full-time employment; [2] unemployed. Freeter often live with their parents as "parasite singles"

Gakusei (学生). Student; connotes lower social status than a worker, cf., *shakai-jin*

Haken-shain (派遣社員). Non-regular (temporary) worker recruited through an agency

Ichinin-mae (一人前). Independent; cf., *jiritsu* and *shakai-jin*

Ii-mawashi (言い回し). Ways of speaking

Ippan-shoku (一般職). The clerical track

Jichō (次長). Vice manager

Jiritsu (自立). Becoming independent in a psychological and socio-economic sense, cf., *shakai-jin*

Jūjitsu (充実). [1] A sense of fulfillment or enrichment; [2] a substantial period of one's life

Jū-yaku (重役). Board of directors

Jōmu (常務). Executive director

Kachō (課長). General section head

Kaikyū (階級). Socio-economic class

Kaisha-in (会社員). A gender-neutral term for office worker

Kaisha kyōdōtai (会社共同体). Firm community

Kaisō (階層). Social stratum

Kakari-in (係員). Workers at non-managerial level

Kakusa (格差). [Social] inequality

Kanri-shoku (管理職). Managerial staff

Karyū-shakai (下流社会). A comment on the growth of social disparity since the 1980s that by mid-2000 shows a downward trend of individuals in the middle stratum of society moving into lower middle and lower strata

Kigyō-shakai (企業社会). Enterprise society

Kigyō-chūshin-shakai (企業中心社会). Corporate-centered society

Kigyōmei-kōhyō-seido (企業名公表制度). Policy that regulates company misconduct through threat of public disclosure

Kinshi-kitei (禁止規定). Regulation enforced by prohibition

Kōhai (後輩). A junior worker in terms of age and tenure within a pattern of vertical social relations

Kone (コネ). An English loan word, shortened form of "connection." In the business context indicates family connections utilized in gaining employment, cf., *enko*

Konpa (コンパ). An English loan word, shortened from "company." Designates a casual social gathering between members of a similar age group. The term is used largely among students but also by company workers. *Gōkon* (合コン) indicates a *konpa* between opposite sex members arranged to facilitate romantic encounters. Typically *konpa* and *gōkon* are used interchangeably.

Kōsu-betsu-koyō-seido (コース別雇用制度). Career tracking system

Ma (間). A space between: [1] a space in the city, free of meaning and function; [2] a space of connectedness in social interaction that is given form and meaning through practice.

Manzai (漫才). Stand-up comedy typical of the Osaka region; there are two roles— *boke* (air-head) and *tsukkomi* (one who corrects the boke's errors)

Mibun (身分). Social status

Midare (乱れ). Disorder

Neet (*NEET*). The segment of the population Not currently engaged in Employment, Education, or Training. Indicates an unwillingness to enter employment and rejection of the socially responsible role of working adult. Increased numbers of *neet* is due to economic stagnation and high rate of unemployment of the 1990s.

Nenkō-seido (年功制度). Seniority wage and promotion system

O-botchama (おぼっちゃま). Preppy boy with a family fortune; variation on *botchan*; also see novel of that name by Natsume Soseki

O eru (OL). Current term for female office workers/ladies that appeared in the 1960s

Ogori (奢り). A social gesture of pride

O-jōsan (お嬢さん). [1]The daughter of so-and-so; [2] young women from sheltered and rich backgrounds

Omote (表). Front; face; appearances; surface. Cf., *ura*

Opun gata (オープン型). Open type: Following the diversification of employment practices, a new type of career track option available for women in the financial sector (c. 2006), like male workers the position involves transfers to any of the company's domestic subsidiaries.

Osana-najimi (幼なじみ). Childhood friend

O-tomodachi (お友達). Friends: [1] platonic relations; [2] in certain usage the prefix *o*-connotes sexuality or ambiguity regarding the status of the relationship

Salary man (サラリーマン). Male white and blue-collar workers earning a salary

Sei-shain (正社員). Full-time employee

Seken (世間). A relational term with a spatial reference. A body of people or workers, who fall between those known to oneself, often identified as close, and those who are strangers. There is no singular or set way of identifying who fits the position of *seken* at any one time. Thus, the relation between the self and *seken* is ambiguous because the boundaries of the term are flexible, relatively arbitrary and dependent on context. It is a Japanese idea of the self that faces inescapable and constant surveillance from those identified as constituting the *seken* category.

Senpai (先輩). Senior worker in terms of age and tenure in a pattern of vertical social relations

Shakai-benkyō (社会勉強). Learning about society through experience

Shakai-jin (社会人). A full-fledged member of society, or a new graduate who has entered the world of work or stage of adult life

Shakai-teki-sekimu (社会的責務). Obligation to society

Shisuta (シスター). From the English word "sister." Describes the supervisory relationship between senior and junior female colleagues. The senior tutors the junior with regard to accomplishing specific tasks and knowledge of the general aspects of working in a business environment

Shisuta seido (シスター制度). The sister tutelage system

Shunin (主任). Junior section leader

Shuseki (首席). Official title of a managing director of an overseas subsidiary

Shūshinkoyō-seido (終身雇用制度). Lifetime employment system

Shusse-kōsu (出世コース). Promotion course

Sōgō-shoku (総合職). The career track

Teiki-saiyō (定期採用). Recruitment procedure in companies to find candidates who commence work from April

Tō-mawashi (遠回し). Periphrasis

Torishimari-yaku (取締役). Director and board member

Tsukiai (付き合い). Indicates an individual's personal relationships ranging from formal and distanced to informal and close

Ura (裏). Back; beneath the surface; mind. Cf., *omote*

Wa (和). Harmony or unity

Yaku-shoku (役職). Rank

Notes

1 Knowledge and Competing Discourses in Organizations

1. Anthropologists of organizations and workplaces will find the texts produced by firms and individuals to be a rich material source for analysis. For an overview of textual analysis in anthropology and linguistics, see Hanks (1989).
2. I describe but do not include the photographs featured in the corporate brochure as I wish to protect the anonymity of my field site.
3. Nonaka and Takeuchi (1995) explore knowledge creation and the significance of tacit knowledge to the innovative capacity of Japanese organizations. The description of knowledge below sounds familiar to anthropologists: Tacit knowledge is difficult to communicate, share, or formalize; highly personal; intuitive; embodied—deeply rooted in action and experience, as well as in ideals, values, and emotions (p. 8). It sounds like culture. However, Nonaka and Takeuchi distinguish knowledge from culture; they argue that studies of organizational culture focus too little on knowledge; ignore the potential and creativity of humans; and the organization is portrayed as a passive entity in relation to its environment, in that it cannot change nor create (p. 42). They are correct to point this out, but we should also note that organization studies have applied the culture concept differently to anthropologists studying organizations (Wright 1994), and what anthropologists study is culture as knowledge and knowledge as culture. The workers who possess this knowledge are central to the anthropological project. In understanding workplaces, the encounter between ideology and other kinds of knowledge has been of central concern for anthropologists.
4. I follow Norman Long's (1996) outline but only note the points that engage this book.
5. I cite from Tomiko Yoda's (2006: 16–53) brilliant overview of this time.
6. The average age at first marriage has been rising: men—29.1 years old; women—27.4 years old (Sōrifu 2004). Between 2000 and 2005 the numbers of unmarried mothers and divorced mothers increased by 142,000 women, particularly among 30–40 year olds, to a total of 1,180,000 single mothers (Asahi Shinbun 22/12/06). Parasite-singles (coined by Masahiro Yamada) contribute to the declining birth rate and economic recession as they tend to be unemployed. Their numbers increased rapidly in the late 1980s. In 2004, parasite-singles

account for 45 percent of the population aged 20–34, and middle-aged parasite-singles aged 35–44 account for 10 percent of this age group (Statistical Research and Training Institute 2006).

7. Osawa Mari (1994), citing Watanabe Osamu, writes that corporate-centered society applied only to regular employees of large private sector firms during the postwar high-growth period until the oil crisis in 1973 (p.160). Companyism was criticized by labor unions of small and medium-sized firms, as well as public sector firms, as an outdated practice holding Japanese society back (p. 162). By the early 1970s the government focused on the welfare state. However, corporate-centered society and the ideology of corporatism was resuscitated following the 1973 oil crisis, amidst slow growth, and the corporate discourse of efficiency and rationalization permeated society (p. 163). Workers in large private sector firms were affected severely, but workers in small companies tied into large firms through the keiretsu system suffered the most; during this time most workers began to identify strongly with the firms they worked for (ibid.). What also emerged was a Japanese-type welfare society in which home and company was expected to sustain the lives of individuals, and caring for the elderly fell to women of the household; thus women were discouraged from taking jobs outside of the home (pp. 170–1). The gendered division of labor and the segregation of labor markets by gender is ingrained in corporate-centered society, and this is engineered heavily by the government through its social security system (p. 173). Patriarchy is thus inseparable from corporatism.

8. For critiques of Pierre Bourdieu's theory, see Adkins and Skeggs eds. (2004); Calhoun et al. eds. (1993); Shusterman ed. (1999).

9. Roland Barthes describes ideology as discourse, termed *ideosphere*, as a total linguistic system: it is only language (2005: 86–93). There can be many coexisting ideospheres, such as political or religious. To paraphrase, the ideosphere succeeds in constituting external causes into interior concerns; it traps or recycles the subject within a space of language, like gum stuck to the bottom of a shoe; the subject feels and sees the ideosphere as a state of mind as it posits a relational distance between subject and language; it has a phantasmagoric quality; subjects within the discourse experience it as universal and natural, it is discourse-law that isn't perceived as law, it keeps power between optimal poles, it crosses boundaries with dangers to itself; it is mimicked, consciously or unconsciously, which poses the question of sincerity; it has the power to endure, which symbolizes the hardness of power, a repetition of monologue; and the ideosphere perpetually speaks inside the subject's head.

2   Fieldwork and Methodology

1. The finer schedule of my fieldwork is listed in Appendix 1.

2. Augustin Berque (1997: 62–6) contrasts Japan and Europe in that each has a different relation of the gaze to the social space of the urban environment: Paris is a " 'space to be seen,' rather than to be lived in" while the Japanese urban landscape is "severed from the aesthetic" (citing Bel [1980]), which explains the absence of descriptive accounts of the urban city in Japanese novels (citing

Takeo [1988: 142–5]) and similarly, a survey of residents of Osaka showed that urban space is to be lived than to be seen (citing Kunihiro [1988]).

3. For similar difficulties of access, see Kondo 1990a; Matsunaga 1995; Turner 1995; and Rohlen 1974.

4. These points concerning issues of access, recognition, and understanding of anthropological research methods by the host organization highlight the contingencies and challenges involved in doing fieldwork in organizations. Such methodological issues are ongoing considerations throughout the duration of fieldwork that have critical implications for writing-up (Hirsch and Gellner 2001).

5. See Lila Abu-Lughod's (1986) account of her father's influence on her introduction to her fieldwork site.

6. This paragraph borrows from the following guides to fieldwork by Atkinson (2007); Augé and Colleyn (2006); Davies (2008); Emerson et al. (1995); and, Hammersley and Atkinson (1995).

7. At JCars, despite the familiarity and warmth that was extended to me from the beginning, I felt self-conscious and awkward in the Japanese corporate environment. As a result of my parents' devotion to catechizing Japanese cultural values and from living in Japan between ages 11 and 15, I was familiar with standard behavioral norms, but the corporation was a world far-removed from what I'd previously known. I am bilingual, also fluent in both Tokyo and Osaka dialects, but had insufficient mastery of honorific language (the different forms are polite-*teineigo*; respectful-*sonkeigo*; humble-*kenjōgo*) and I worried constantly. It was also my first fieldwork assignment. At the time I was 23–24 years old. By nature, I am easy-going and pacific. My transnational upbringing colors my identity: I was born in Japan but uprooted before any conscious identification with Japan was made so, between ages 4 and 10, I happily assimilated to a middle-class North American childhood in Rocky River, OH, the mid-West. I lived in Osaka and Nara between ages 11 and 15, and by the time I had come to accept my uneasy place in Japanese society, my family relocated to England. I have resided in London since attending university at age 18. Meanwhile my parents left the UK, resuming their peripatetic ways in Japan, on to the United States, France, then back to Japan. I was educated at state schools, with the exception of attending a single-sex private school in England. While I identify strongly with the bohemian intellectual's cultural landscape of London and Western Europe, a "Japan" of my family and early associations reaches out to me in its absence.

8. Here I draw on Marc Augé's writing on otherness where he makes the distinction between ambivalence and ambiguity (1998: 30–1).

3 Gender Segregation and the Japanese Labor Market: Equal Employment Opportunity Law

1. The American equivalent to the salary man is the "organization man" or the corporate man who not only works for, but belongs to the organization or collective, removed from the ideal/mythology of individualism instilled in the Protestant work ethic (Whyte 1956).

2. "When media or politicians question the security of U.S.A. foreign investment, all attention turns to Japan, despite the fact that British and Dutch nations are equally significant owners of U.S. assets... Japanese trade practices are often called 'predatory' capitalism or trade practices, but very little evidence suggests that Japanese companies coming to the U.S.A. is any different from behaviour of the U.S.A. going into Europe after the war... Predatory practice refers to a company or country targeting a sector, industry or a niche... [However,] the basic axiom of capitalism is that growth is contingent upon the expansion of markets and all countries in this light will find reason to come to the U.S.A.' (Hettinger and Tooley 1994: 37–9). In addition, idioms of racial stereotyping of "the Japanese" commonly found in the contemporary United States forms the basis of Japan-US political tension (Dower 1986; 1993).

3. NB. Figs. for United States and UK are for 16 years old and above; for Japan over 15 years old. Source: International Labour Organisation, 2005.

4. By status I mean to describe a certain social condition experienced by an individual through distinctions by rank (occupation, age, sex, etc.) that might be experienced as a group and by distinctions within the group. Status is continually capable of shifting as it is acquired through negotiation and interaction in the workplace.

5. For complementary cross-cultural ethnographic analyses of women's power in the domestic sphere, see Gullestad (1984) and Wolf (1972).

6. Women in older cohorts might be experienced and skilled in their jobs, but if they are assigned to the clerical track they receive lower wages and status compared to the inexperienced newcomer in the career track. This was a source of resentment for women of older cohorts within the clerical track (Takenobu 1994: 43). Conversely, women who enter the career track might compare themselves to female graduates of junior colleges who have a head start in the work environment. The older woman entering the company through the career track might compare herself negatively to them, evaluating her inexperience on the job, relative to her age and to the age of other women younger than her as a reflection of incompetence. Interpersonal relations between part-time workers (mainly older housewives) and full-time workers (younger female university graduates) in a Japanese chain store exhibit similar complexities based on age and status. Part-timers with higher status derived from age and length of service were reluctant to credit the full-timers who were superior only in terms of official rank and status (Matsunaga 1995: 302).

7. Women could thus close the gap in gender inequality in the workplace as men cease to exercise paternal protection over women; in other words, protection has been a means through which men have constrained women (Meillassoux 1981).

8. This difficulty is compounded by the government-business "Old Boy" network, consisting of ex-government officials re-employed by the private sector as board members (Johnson 1982; Nakane 1970; Schaede 1995; van Wolferen 1989).

9. *Seken* indicates the relation between the self and other. Discussed in chapter 8.

10. This seems to indicate that relations between wife and husband have become closer in the salary man household: in the past 20 years, male participation in

housework increased on average by 6 minutes a day, but by 21 minutes on Saturday and 25 minutes on Sunday (*Sōrifu* 1996). This emphasis on home life by salary men might indicate a reversal of traditional notions of Japanese marriage where companionship was relatively unimportant (McKinstry and McKinstry 1991: 31).

11. Company restructuring in Japan differs from the timely and effective method taken by American companies. Japanese companies consider the re-employment opportunities of the individual and aim for what is termed a "softlanding" type restructuring that involves outplacement (the individual is sent to a small subsidiary to acquire new skills, then returns to the parent firm, and goes back to the subsidiary for a permanent post) and outsourcing (*Nikkei* 19/08/98). Recruitment companies such as *Pasona* support this type of restructuring by registering middle-aged men; *Pasona* focuses specifically on marketable skills to increase their candidate's chances of re-employment. This re-employment is taking a new form: men are increasingly working from home by utilizing personal and internet media (*NSS* 19/08/98).

12. The total number of divorcees has been increasing in every age cohort of the population, but sharp increases are seen among 20–29-year olds. This trend is termed *Narita Rikon* by the media as women leave their husbands at Narita Airport on returning from their honeymoons. Women in their fifties have been increasingly divorcing husbands as they retire, which reflects the difficulty of adjusting to the presence of their husbands in the home over long periods. These trends point to the growing independence of women and their disillusionment with the patriarchal institution of marriage.

13. For examples of restructuring in other firms see Asahi Corporation (*NSS* 07/08/98b); Nikkei Kin and Shinnitetsu (*Nikkei* 01/08/98); Mikuni (*NSS* 03/08/98); and MITI plans to restore the manufacturing industry (*NSS* 16/09/98).

14. Clark (1979) and Nakane (1970) had predicted the likelihood of these changes.

15. In his study of workers in a Tokyo bank, Thomas Rohlen (1974) assigns women lower status despite their occupational role because they are women (also see Nakane 1970: 33, n.*). This could be related to his understanding of status as being a supervisory power.

4 Firm Entry, Tracking System, Careers, Status Negotiation

1. Interview with Ashikawa-*jichō*, my section manager in the overseas administration department. He works closely with the management administration department.

2. Rohlen (1974: 68) notes that women are more likely to be recruited by connections because they will not stay employed at the bank for long. Generally, access to jobs via personal networks is not necessarily gender-based or a process that is unique to Japanese firms; the role of these networks is theorized in the literature on social capital in labor markets (Lin et al. eds. 2001). Beyond an individual's personal networks, Mary Brinton (2000) analyzes ties between Japanese high schools and employers as institutional social capital.

3. During the economic boom, JCars' recruitment process of junior college graduates is more varied. After obtaining the consent of the candidate, the junior college nominates candidates who suit the requirements and preferences of a company (*suisen-seido*). The company and the college have a no-loss agreement for this type of recruitment. If the company offers a position to a candidate, the candidate is unable to refuse. This system enables the company to give out the exact number of offers for the positions to be filled.

4. I do not discuss matters of cost, since this is confidential information. However, this should not affect the general thrust of my argument.

5. Japan Organization for Employment of the Elderly and Persons with Disabilities (JEED) is an independent administrator, certified by the Ministry of Labour, promoting the civil rights of elderly and disabled people in line with government legislation. Alongside Japan Association for Employment of the Disabled, JEED offers professional expertise concerning the employment, training, and integration of disabled persons: they mediate between the employer and disabled employee as "job coaches," and instate teams consisting of representatives of disabled and non-disabled workers in order to establish disabled employees in their work environment.

6. Other temporary workers in the company work on the manufacturing lines in the plants located in the neighboring Prefecture. Temporary workers in the plant are of two types. First, workers are recruited from other companies, in which the company paid the workers via the company at which they are officially employed (*shagai-kō*). Second, workers employed by other small manufacturing firms are invited to work at the company (*ukeoi-kō*) to use the facilities to manufacture the particular parts for the company.

7. Transfers occur across sections and departments, whereas job rotations occur within the same section.

8. See Ogasawara (1998: 47) for a similar account of female workers unable to unite as a result of status differentiation introduced among women by the EEOL. Lo (1990) and Rohlen (1974) also observed how the high turnover of women functions to dilute social bonds.

9. Corporate norms in the 1970s as described below still shape current workplaces and individual careers, but socio-economic shifts introduce variation to form and this is my point. "[W]hat 'lifetime employment' there is could be construed as the effect of a labour market consisting of essentially self-serving individuals and firms [lifetime employment is applicable exclusively to large firms]. But this sort of explanation does not invalidate the proposition that 'lifetime employment' is at the same time an ideal, and a very powerful one, entailing an obligation of mutual attachment between firm and employee. Sanctioned by what is seen as tradition, morally correct, and emblematic of Japanese culture, 'lifetime employment' is the goal towards which both firms and individuals have to direct their efforts—or their apologies" (Clark 1979: 175). Also see Cole (1979: 20) for a similar point.

10. As noted by many commentators, the body is socially and culturally produced and therefore culturally specific. "The physical experience of the body, always modified by the social categories through which it is known, sustains a particular view of society. There is a continual exchange of meanings between the

two kind of bodily experience so that each reinforces the categories of the other" (Douglas 1996: 69). People invest both economically and psychically in the body (Baudrillard 1998) and the body, beauty, and femininity become markers of social status or capital that mark out class relations (Bourdieu 1984; Skeggs 1998). Appearance and beauty is an aspect of patriarchal, corporate, and capitalist ideology that constrains women both western (Greer 1991; Wolf 1990) and Japanese (Komashaku ed. 1985; Osawa, H. 2000), and increasingly men (Riley 2005: 30–47; Wolf 1990). Watanabe Tsuneo (1995: 122–6) writes that beauty belongs to women and cannot be claimed by men. Indeed, the pervasive and bland salary man body is not seen. It is even precluded from the objectification that Simone de Beauvoir says reduces women to a passive subject, but, Watanabe goes on to say that without being seen the salary man cannot claim the objectification necessary in order for the body to emit non-verbal cues. From this perspective, salary men envy women whose bodies are more than machines that walk to work, feed the brain at work, and finish up as a heavy load at the end of the work day. Male beauty, however, defines contemporary masculinity as Laura Miller (2003) observes; in fact, it caters to female desires that oppose the salary man body.

In workplaces, appearance, masculinity, and femininity, are standards that evaluate employees. The body engages with the world, and appearance marks the embodiment of cultural values that enable evaluations of individuals because it acts as an objective marker of difference (Bourdieu 1984: 192). Appearance is important for individuals as it is intrinsically linked to one's aesthetic and moral values (p. 206) and where there are material and symbolic gains, particularly in the workplace, appearance even takes on "occupational value" (p. 202).

Linda McDowell (1997) takes up the connection between appearance and success in workplaces in her study of gender segregation in London's financial district. She argues that City workers are socially constructed through discursive practices that organize gendered and embodied performances. Masculinities and femininities are thus enacted in multiple ways within the parameters of a dominant discourse—identities are fluid and negotiable, although they tend to reproduce gender segregation.

Romit Dasgupta (2000) locates the creation of the salary man as the hegemonic masculine ideal in context of patriarchal industrial-capitalism. He shows how individual men negotiate identity in relation to a carefully constructed and powerful discourse, displaying conformity at many levels—model of life, life style, and everyday behavior and appearance as the ideal worker. Yet, ambivalent attitudes—accepting, cynical, and resigned—characterize the worker's relation to the hegemonic discourse. Following Judith Butler's idea of performativity in identity construction, Dasgupta describes the "crafting" of men to fit the salary man ideal, which peaks when they enter the workplace. He also notes changes in the cultural image of the salary man following the postbubble recession—weak and pitiable, salary man running away, escaping. Brian McVeigh (1997) explores how sociopolitical and economic forces construct and define bodily management among junior college students.

11. A related theme emerges from Painter's (1993: 307) account of ideology and popular culture based on fieldwork at a television company in Osaka: "Producers and directors in Japan agree that it is a mistake to include too many outspoken *tarento* [television star/personality] on a panel: the ideal balance is one that allows space for followers as well as leaders... [for] the creation... of consensus."

12. My general observation is that during work situations most workers are demure, giving the impression of homogeneity, but at moments throughout the working day, during lunch hours, and in after-hours workers exhibit exuberance and demonstrate their various personalities.

13. A suitable appearance at interviews requires that men have short hair showing the forehead and ears; women with long hair tie it back; women wear skin-colored tights. Giving an impression of cleanliness or freshness is key (*Nikkei* 30/01/98).

14. A curious article linking facial features to the recruit's attitudes and motivations suggests that a larger nose connotes ambition to become CEO, voluminous hair indicates a tendency to stay at one company throughout his career (*Rōseijihō* 1997: 43).

15. Department stores arrange a special recruitment-ware section from December. Most suits are navy, but grey is also a popular color for women's suits (*Nikkei* 30/01/98).

16. Women have questioned the need for uniforms in offices (understandable in laboratories or factories where clothes get dirty) as a form of gender segregation, and companies have started to abolish uniforms that are only given to women, but this is not a victory for gender relations, rather it is a way for the company to cut costs (*Nikkei* 21/09/98). The wearing of uniforms is also an intersex issue: status differences between women are expressed at Toyota by women in the career track wearing mufti whereas women in the clerical track are issued uniforms (Ashimoto-*shunin*).

17. My impression is that the Japanese notion of the body and beauty is closer to the French than Anglo-American.

18. Social inequality (*kakusa*) is said to have widened, and Japanese social structure is shifting away from a largely middle-stratum to a class society like the United States and UK. This underlying structure of society marked by greater inequality is termed *kakusa-shakai*. To simplify, between the mid-1990s and mid noughties, the upper class increased marginally, the middle class fell, and the lower middle and lower classes increased. More specifically, Miura Atsushi (2005) calls this society, with the fall of the middle and increase in the lower middle band, *karyū-shakai*. The finer differentiation within classes and a shift in the growth of the new lower-middle class has roots in the 1980s high growth period. As the economic and population structure changed (prolonged recession; lower population and birth rates; inequitable redistribution of wealth) in the 1990s, this was mirrored in changes in employment structures and labor markets (end of lifetime employment; growth in non-regular forms of labor), family structure (fewer marriages) and education; and a diminished sense of hope for a better future characterizes emotions within *karyū-shakai*.

19. Independence (*jiritsu*) as Amano Masako (2002: 252–3) explains is a concept rooted in relations between self, others, and society. The term is historically attributed to men, where independence was achieved in relation to the nation-state and society. In relation to women, first-wave feminists critiqued institutional barriers to women's independence, and second-wave feminists reviewed personal relations within the home—reproduction and marriage; subsequently, a woman's right to make choices became attributed to the individual rather than the nation-state. Then issues of women's power in decision making were taken up in the context of workplaces and occupations, financial and spiritual domains, and in relation to civic life. A complimentary notion of adulthood is noted by Louella Matsunaga (1995: 218): "*ichinin-mae* (一人前) indicates a...certain level of maturity and personal development [which is] constrained by gender, in which both work and position in the household play a role."

20. Mikhail Bakhtin's ideas have gained a massive following in the Western humanities and social sciences since the late 1960s. The Bakhtinian project that houses the plurality of multiple, mini narratives was developed as an alternative to the grand narrative of communism in the Soviet Union; this is why his ideas ally with postmodernist and post-structuralist approaches that are favored by analysts today (Eagleton 2007). His method politicizes the nature of social and historical conflict, rescuing conflict from the status of static to active force (Crowley 2001: 179); privileging agency over structure (Aronowitz 1995: 121) while the word that circulates is for all participating agents ("speaker, writer, listener, reader") "a contextually embedded, socially constituted, intersubjective event that allows for unfinalized, but not indeterminate, meaning" (p. 91). Indeed, Bakhtin points out that "language is not a neutral medium" and the polyphonic character of literary discourse embodies the belief that "almost no word is without its intense sideward glance at someone else's word" (Peterson 1995: 94, 97).

Nancy Glazener (2001) observes that while Bakhtin himself did not address feminism or gender differences and their social affects, his approach accommodates gender as a social and discursive category (p. 155). She revises Bakhtin's notions of subversion and the carnivalesque for use in her literary analysis of the novel.

Among anthropologists Bakhtin's ideas have gained wide application: Jane Hill (1986: 94) writes that a Bakhtinian approach is relevant to the anthropology of language in relation to "whether anthropology can define a notion of the human self that will allow the voices of others to speak without submerging their uniqueness within each culture and historical situation, and the question of how the voice of the anthropologist should engage the voices of these other selves."

Wendy Weiss (1990) points out that Bakhtin's framework entails recognition of the value and distinctiveness of the subject's experience, and importantly of all parties present in the dialogue. She raises the utility of Bakhtin's approach to ethnographic description and theoretical examination of social action in a study of gender relations in Ecuador: Weiss finds that not only is ideology reflected and reiterated, but dominant positions are caricatured and subverted (p. 424). Weiss picks up from Bakhtin's work that intelligence is returned to the subject (p. 422–3). To this I would add dignity. She also directs our attention to the similarity between Bakhtin's oeuvre with postmodern critiques of ethnographic authority of the 1980s.

And Vincent Crapanzano (1995) challenges postmodernism's claim that it posits a decentered narrative of the world. By drawing on Bakhtin's ideas of parody and stylization, two passive types of double-voiced word, Crapanzano identifies inequalities that structure all cross-cultural dialogues, this, he argues, is always overseen by an authoritative Third that mediates all linguistic exchanges.

5 Linguistic Spaces of Vertical and Horizontal Organization

1. This egalitarian counterpart is necessary as her aim, situated historically, is to defend the tradition of the Japanese system against domestic and international critics who espoused modernization and labeled Japanese social organization as "feudal" (Nakane 1970: 66).

2. There is a pejorative word that describes OL as *soshiki-nai-geisha* (geisha in organizations). The use evokes the image of female office workers pandering to men's needs. Lest there be any confusion, I wish to state that my comparison does not subscribe to this sense of womanhood.

3. The competitive, catty side of relations between geisha is depicted comprehensively in Arthur Golden's (1997) historical novel, *Memoirs of a Geisha*. For a critique of Golden's Orientalist rendering of geisha, see Allison (2001).

4. As stated previously, in Japan sibling relations are considered hierarchical, not lateral units; what, then, are we to make of the comparisons and rivalries that resemble more the feelings that emerge within relations between individuals in the same category?

5. Women and men comprise the *dōki* group, and members are identified by the application of the informal suffix; *kun* (masculine) or *chan* (feminine); or by nicknames. Among men, at times, depending on the context, when referring to each other the suffix is removed.

6 Temporal Dimensions of Symbolic Community

1. Some Japanese companies are very strict with regard to punctuality: "I remember having long commutes to work when I was in Japan. One time I had to commute for two hours to Odawara from the heart of Yokohama. I was five minutes late, because of the trains, and I had £15 docked off my pay, no questions, no nothing, the payroll computers did it automatically, on company policy" (Lucy, my London-based friend).

2. Jack, an American language teacher, notices similar "doing nothing" behavior among male colleagues at his junior high school in Nara:

    The male teachers at my school are always smoking and stay late, until 8PM, with nothing to do. They don't seem to be getting on with much, and they say to each other: "What time are you staying till?" or "What have you got to do?" and the reply is always, "Nothing much."

    I think they feel bad if they leave before the others. They can't say: "Oh well, I am leaving at 5PM today," because, it just looks bad. Maybe they're just inefficient workers, but there isn't much work to warrant being inefficient. If they spent less time in the smoking section where they go to smoke, they might be able to go home earlier.

3. In this context of drinking as office time, women are also invited, albeit less frequently than men and particularly when managers of overseas subsidiaries return to the head office for business trips.

4. The quality of the relationship between the manager and junior member is important to an individual's promotional prospects, particularly in large Japanese companies where promotion and wage levels are characterized by strong internal labor markets.

5. The attention to the quality of *ningen-kankei* in the senior-junior relationship works as much in reverse. Nakane (1970) notes: "More than anything else, the qualification of the leader in Japanese society depends upon his ability to understand and attract his men. No matter how great his wealth and power, how brilliant his talent or what type his personality, if a man is unable to capture his followers emotionally and glue them to him in vertical relationships he cannot become a leader" (pp. 73–4). Furthermore, "[t]here is a high degree of personal involvement between manager and employee" where the former attends the latter's wedding, or in reverse, in the case of funerals within the director's family, employees take charge of arrangements (p. 76). Rohlen (1974) also notes that the style of the manager is important when promoting solidarity within the section (pp. 116–7).

6. Jeannie Lo (1990, chap. 4) describes the parental role of dormitories in the lives of female workers whose parental homes are too far for commuting.

7. As a general sociological trend, since 1985, the structural mobility in occupational categories between the father's generation to the son's has shown upward trajectories allowing mobility from blue-collar to white-collar jobs in this generation (Seiyama 1994: 54–7). Therefore, Kubo's experience is not unusual in a structural sense, but it might be unusual from the perspective of social-class reproduction, assuming that family background, class, and personal decisions that are themselves related to class socialization and enculturation, contribute to the decision making of individuals with regard to the choice of schooling and careers (Roberson 1995: 304).

8. Oë Kenzaburo, the Nobel Prize winner for Literature in 1994, writes about disability in his many novels, and in his acceptance speech describes the experience of fathering a mentally handicapped son in an intolerant society (Oë 1995).

9. The study of food is a major subdiscipline in anthropology, see Mintz and Du Bois (2002).

10. As argued in this book, I do not picture management discourse and workers' practice as oppositional categories.

7 Spatial Practices and Hierarchy

1. Ideas of time-space compression (Harvey 1989) and time-space distanciation (Giddens 1990) have been central to the way geographers and sociologists understand modernity. Time leaves its traces but in a present space, and thus

time-space are two inseparable aspects, not two inseparable ideas (Lefebvre 1991: 37). Space influences social practices and, equally, social practice constructs and gives meaning to space. Anthropology is concerned with the symbolic meanings of space, and focuses on the relationships between daily routines, linguistic discourses, power relations and social structures and embodiment (Moore 1986).

Various types of space exist: somatic space, that accounts for the space of the body framed in conscious action; perceptual space, that is, a space created around individuals by their intentions and behavior; existential space, which is a social level in the understanding of space; and architectural space, which considers the interrelations between a built space and the types of space given above. Spatial topics that have engaged geographers, anthropologists, and cultural studies scholars include the city and architectural formations in connection to social relations (Berque 1997; de Certeau 1984; McDowell 1997; Soja 1989); the politics of representation (Bhabha 1994); metaphorical space (Foucault 1977); connection between space and micro-physics of power (Foucault 1972); and capitalism and control (Ong 1987).

There is interesting work on space with respect to Japanese studies: Kondo (1990a: 57–75) accounts for the symbolic correlation between socio-economic class and geographical divisions in the space of Tokyo city, particularly in connection to the respective identities reflected in the division of uptown (Yamanote) and downtown residents. Kondo then takes us into her sweet factory, exploring its spatial dimensions in relation to inside and outside (*uchi/soto*) as the background for the production of meanings about family and the crafting of selves. She writes that the household and inside (*ie* and *uchi*) "structure language, behaviour, space, and feeling...[and] define a world and give its members a place in that world...Yet [household and inside] cannot mean the same things to all people" (p.159); while Bestor (1989) looks at the spatial class boundaries within a Tokyo ward through practices of place-making.

Berque (1997), among other points, accounts for the structure and patterns of usage of metropolitan space. Berque refers to the Japanese spatial concept of *ma* that has a double meaning: first, as a free space in the city, the interval in the cityscape—a wasteland, free of meaning and function; and second, as a term describing human social interaction (*ningen-kankei*) that is a space formed by and given meaning through action, thus, implying a space of connectedness (p. 141). Further, he discusses a specific construal of space (*ba*), where the experience of the social becomes intertwined with myth; until the 1940s, the collective identity of the Japanese as a national body (*kokutai*) was synonymous with the emperor, as the emperor's body/person was regarded as a certain site (*basho*) (pp. 135–6). Berque also writes about the fluidity of space through reference to movable partitions with Japanese homes.

Lebra (1992a; 1993) explores the upper class Japanese household, and the structuring of space in relation to gender and status between the aristocrats and their servants. In short, Lebra (1992a) discusses three spatial dimensions of interaction that apply to the interrelationship between people that become expressed in space: top/bottom (*ue/shimo*); front/back (*omote/ura*); and inside and outside (*uchi/soto*), and an individual experiences either one of these dichotomous positions in all three spatial dimensions at once. For example, an upper class female occupies the top, back, and outside on account of status and gender, whereas an upper class male occupies top, front, and inside in the home. Both male and female aristocrats occupy positions at the top of the house and in contrast, the male and female servants occupy lower positions. Thus, Lebra shows that the gendered hierarchy does not follow a straightforward vertical model. Instead status difference is expressed in the opposition between back/front and inside/outside.

Tobin (1992) examines the way preschools inculcate cultural values through symbolic meanings associated with transitional spaces. (For meanings of transitional space on a global scale see Augé 1995.) Tobin (1992) writes that Japanese buildings have a designated area near the door where outdoor shoes are removed before entering, and the child learns to perceive his or her identity in relation to the preschool group through this custom of changing into indoor shoes. The ritual practice of greeting and parting that takes place within this transitional space is crucial to imparting the distinction between inside and outside (Hendry 1995: 44).

Referring to two "scenes" relating the interaction taking place between members in different areas within a household, Bachnik (1998) argues that relations between *omote/ura* have been misconstrued; they are not dichotomous, instead they are terms signifying the distance between individuals in the construal of self and other, which differ according to the degree of formality defining the encounter. Thus, it is the degree of formality that positions the self in relation to others, not inside/outside. Bachnik, thereby, engaging with critics of the group model Befu (1980) and Mouer and Sugimoto (1980), questions the holism of models given to explain the self in interaction that reduce the social to abstract terms. Essentially, as Barthes (1982), Doi (1973), and Hamabata (1983) have argued (cited in Bachnik 1998: 108), the anthropological terminology describing a person in relation to others, *omote/ura*, and its associated terms *tatemae/hon-ne*, *giri/ninjō* (social obligation/feelings), *soto/uchi*, focuses only on one side of the postulate because, the relation between these terms are taken to be dichotomous. Bachnik's point, then, reinforces the idea that both elements are present in any one social scene, but it is only one that is presented to those positioned as others, for example, houseguests (p. 108). Importantly, this linguistic theorization linking Japanese terms of reference for self/other to practice incorporates the actual space of the household in its analysis.

Traweek's (1988) ethnography of physics labs in Tsukuba, Japan, and Stanford, United States, maps space and its uses, and the divisions and relations within (pp. 18–45). In an ethnography of a Tokyo bank, Rohlen (1974: 105–6) describes the arrangement of desks that are inward facing, as in a circle, so when seated, workers do not look at the backs of other workers of their section. Rohlen (1974) analyzes spatial arrangement as analogous to the quality of interrelationships between workers: the section is harmonious, and vertical ties between the head of the section and workers are "direct and close." In contexts of interaction among group members when outside of the office, this hierarchy changes to a more circular way of relating to others (p. 106).

In Nakane's (1970) construal of space, frame or *ba*, indicates a situational position, describing a particular location with a boundary that gives members of a group a common basis for identification. This implies the existence of a unique relationship between members. Nakane uses a spatial term, but does not develop it.

2. In office settings, diagrammatic representation of seating arrangements indicating status distinctions appear in studies by Hamabata (1990: 13–4), Nakane (1970: 33, 35) and Ogasawara (1998: 22–3).

3. When I began fieldwork, most workers warned me to remember the name and rank of senior workers in conjunction, e.g., X-*kachō*. Failing to get this right would have reflected badly on me and offended them.

4. The section head is distinguishable from the general section head in tenure, age, and seniority. The latter is the overseer of several sections in a department (as opposed to leading a single section); therefore, the level of responsibility given to him is greater.

5. "...[S]ymptoms of tension and discontent were the continual and frequent allegations of sycophancy and favouritism. Men would constantly be accused—though not to their faces—of having risen by flattering (*gomasuri*) their bosses" (Clark 1979: 205).

6. In relation to Foucault (1977), Deleuze argues that the disciplinary society of the eighteenth and nineteenth centuries are being taken over by "control societies" (1995: 177–8). In a control society, individuals pass through various sites of confinement, it is rapid and the mechanism is not tied to one place (p. 178), yet it is "... continuous... whereas discipline was long-term, infinite, and discontinuous" (p. 181).

8 Spatial Practices of Mēru and Bulletin Board

1. Practices within organizations in the information age defy hierarchical social structures. This is a characteristic inherent to the medium of the Internet. This position is an extension of a general hypothesis borrowed from interpretations of social dynamics generated by new forms of interaction between users of the Internet. The Internet disrupts conventional notions of temporal and spatial borders. From the perspective of management, an economy based on information (information as a key to acquiring wealth as opposed to land or machinery) implies changes to the way information is organized and distributed within a company. Thus, in the information age, the structure of the company can no longer maintain the centrality of hierarchy preferred under the classic Weberian bureaucratic model of organizational theory (Weber 1982).

For the organization to survive and succeed, it must adapt to horizontal organizational principles. This is because information has a rhizomic character whereby the flow of and access to information cannot be contained and controlled easily by those in higher positions within hierarchical structures. This suggests that efforts to control the practices of workers in relation to the uses of this new technology will increase, as management attempts to maintain the existing hierarchical structures.

This does not mean, however, that hierarchies will disappear, but they can be altered or be supplanted by another mode of organization that might be more suitable. For example, when reading email, hierarchies can be recreated on different principles, according to personal preference (Jordan 2000: 81). Moreover, the new mode of information exchange produces "new, liquid, and multiple associations between people...and new modes and levels of truly interpersonal communication come into being" (Benedikt 1991: 123).

2. Despite the fragmented nature of human experiences in modernity, as Giddens (1994) argues, we are able to establish communication based on a shared discursive space. Furthermore, Internet space has a physical presence in chips, circuits and cables, as well as a phenomenological existence (Benedikt 1991: 131).

3. "Using the Internet then becomes a process of reading and writing texts, and the ethnographer's job is to develop an understanding of these meanings which underlie and are enacted through these textual practices" (Hine 2000: 50).

4. The Instant Messaging service (IM) enabled by the Internet is a parallel example to email; IM ensures deeper and timely communication between individuals who have established relationships offline. The effects of IM on social relations in the office will not be considered here as controls to access were set, that is, no connection to the World Wide Web from individual's laptop computers.

5. LAN operates on a localized network system specific to a geographical location exclusive to the function of particular departments. LAN is an aspect of Management Information System, a comprehensive system that allows for the transmission of management data, fundamental in the decision making process, to appropriate, timely, and correct destinations (New Encyclopædia of Sociology 1993: 353).

6. Owing to a preference for a visually-oriented culture, during the 1980s Japanese workplaces were slow to adapt to alphabet-based keyboard technology and electronic communication (Low et al. 1999: 133–6).

7. The freedom of the Internet is not suited to the controlled nature of Japanese society, particularly for security-conscious companies for which the Intranet is a better option (Low et al. 1999: 140).

8. A common concern regarding productivity exists in Britain: "[m]illions of emails are created by employees playing office politics by sending out multiple copies to other people who do not really need them (and blind copies to others), or covering their backs by sending copies to managers who do not want or need to read them" (*Guardian* 07/06/00).

9. "Organisations are traditionally built around two key concepts: hierarchical decomposition of goals, tasks, and the stability of employee relationships over time. In the fully networked organisations that may become increasingly common in the future, task structures may be much more flexible and dynamic. Hierarchy will not vanish, but will be augmented by distributed lattices of interconnections." (Sproull and Kiesler 1993: 117) (cited in Jordan 1999: 83).

10. Susan Napier's (1991) account of the wasteland of urban Japan analyzes the portrayal of contemporary society by postwar generation novelists such as Mishima Yukio and Oë Kenzaburo, known for their presentation of alternative

worlds, usually in erotic discourses, to which escape or attack is possible from a fragmented, disappointing, arid reality, populated by smug people who are reluctant to question society (pp. 1–44). Such a "sense of loss" is a recurrent postmodern theme among writers, and 1980s Tokyo is portrayed as a materialistic "morcellized universe of fashionable brand names and one-dimensional characters" (Tanaka 1982, cited in Napier 1991: 13). Similarly, Yoshimoto Banana portrays "being" as experienced by the younger generation in terms of a "vacuous sense of life"; of nothingness and meaninglessness (*sono nani mo na-sa*); where individuals reconstitute the meaning of their lives by giving distinctive interpretations to traditional expectations or ideologies (Treat 1993). See also the work of Murakami Ryu, particularly its emphasis on amoral, dystopic urban life.

11. Perhaps this thinking about what constitutes work and non-work practice is made possible as postmodernity (including Practice Theory and the Internet as an object of study) constitutes the subject as multiple, unstable, and diffuse: it brings us closer to a fresh concept of "work" that has not yet been articulated, although management and workers are both aware of it.

12. In Britain, "[t]wo years ago email users were spending 90 minutes a day on their habit. That has now risen to almost three hours, according to a survey by the User Group, and on present trends could rise to four hours a day in a year's time. At this rate it will not be long before the working day is spent reading and writing emails, leaving no time at all for the business of actually making things" (*Guardian* 07/06/00). Quite! There is a Machiavellian trend frequently observed in the UK: the disingenuously composed "spontaneous" email— intentionally disguised as a rapidly written note, but in fact written quite carefully to convey a self-consciously distinct mood (John Swenson-Wright, personal communication 29/10/08). In the UK, business email tends toward informality; in Japan, corporate emails must observe the formal styles and strict etiquette that traditionally govern the handwritten letter, this includes observation of conventional salutations, references to the seasons, and use of honorific language.

13. The RGB Show (1999) was a performance of interactive media that coincided perfectly to my ruminations about my field data on email romance. Our shared concern is with the relationship between technology and the social. The segment of the performance on which I draw here hangs on the concept of synergy between digital imagery and sound. Antirom challenges conventional perceptions of reality, by creating a vacuum where gesturing in communication is removed, and meaning depends solely on words (Nicolas Roope, personal communication 02/06/99). The concept framing this dialogue is not remote from anthropological concerns: the role of computers in the most intimate aspect of relationships in contemporaneity.

14. Previously published material, Kurihara (2007).

15. The CIA enforced disciplinary action on some of its employees who were using chat rooms on the agency's computer system for exchanging gossip, musings, and observations. Eight employees were exonerated while seventy-nine employees were given warning or security briefings (Reuters 01/12/00).

Conclusions

1. I think of Zygmunt Bauman's (2003) writing on discontinuity as a feature of postmodernity; Karen Kelsky's (2001) transnational imagination, sexuality, and desire; and, Gaston Bachelard's (1994) poetics of space.
2. I return once more to Julia Kristeva's (1991) "foreigner". For me, like many others, a need for a sense of *belonging* captures the reality of living in a deterritorialized world, where our contemporary lives exhibit a "generalised condition of homelessness" whether one moves geographically or not (Gupta and Ferguson 1992).

Selected Bibliography

Abegglen, J.C. 1958. *The Japanese Factory: Aspects of Its Social Organization*. London: Free Press.

Abu-Lughod, L. 1986. *Veiled Sentiments: Honor and Poetry in a Bedouin Society*. Berkeley, LA, London: University of California Press.

Acker, J. 1990. "Hierarchies, Jobs, Bodies: A Theory of Gendered Organizations." *Gender and Society* 4: 139–58.

Adkins, L. 2004. "Reflexivity: Freedom or Habit of Gender?." In *Feminism After Bourdieu*, ed. L. Adkins and B. Skeggs, pp. 191–210. Oxford, UK, and Malden, MA: Blackwell/Sociological Review.

Adkins, L. and B. Skeggs, eds. 2004. *Feminism After Bourdieu*. Oxford, UK, and Malden, MA: Blackwell/Sociological Review.

Allison, A. 1994. *Nightwork: Sexuality, Pleasure and Corporate Masculinity in a Tokyo Hostess Club*. Chicago: University of Chicago Press.

———. 2001. Memoirs of the Orient. *Journal of Japanese Studies* 27 (2): 381–98.

Anderson, B. 1991 [1983]. *Imagined Communities: Reflections on the Origin and Spread of Nationalism*, revised ed. London and New York: Verso.

Appadurai, A. 1988. "Putting Hierarchy in Its Place." *Cultural Anthropology* 3 (1): 36–49.

———. 1990. "Disjuncture and Difference in the Global Cultural Economy." *Theory, Culture and Society* 7: 295–310.

———. 1991. "Global Ethnoscapes: Notes and Queries for a Transnational Anthropology." In *Recapturing Anthropology*, ed. R. Fox, pp. 191–210. Santa Fe, NM: School of American Research Press.

———. 1996. *Modernity at Large: Cultural Dimensions of Globalization*. Minneapolis: University of Minnesota Press.

Applbaum, K.D. 1995. "Marriage with the Proper Stranger: Arranged Marriage in Metropolitan Japan." *Ethnology* 34 (1): 37–51.

Apter, T. 2007. *The Sister Knot: Why We Fight, Why We're Jealous, and Why We'll Love Each Other No Matter What*. New York and London: W.W. Norton.

Aronowitz, S. 1995. "Literature as Social Knowledge: Mikhail Bakhtin and the Reemergence of the Human Sciences." In *Bakhtin in Contexts: Across the Disciplines*, ed. Mandelker, A. and C. Emerson, pp. 119–35. Illinois: Northwestern University Press.

Asad, T. 1979. "Anthropology and the Analysis of Ideology." *Man* 14: 607–27.

———. 1993. "Toward a Genealogy of the Concept of Ritual." In *Genealogies of Religion: Discipline and Reasons of Power in Christianity and Islam*, pp. 55–82. Baltimore and London: Johns Hopkins University Press.

Atkinson, P. 2007 [2001]. *Handbook of Ethnography*. London: Sage.

Atsumi, R. 1980. "Patterns of Personal Relationships: A Key to Understanding Japanese Thought and Behavior." *Social Analysis*, Special Issue, Japanese Society: Reappraisals and New Directions, No. 5/6: 63–78.

Augé, M. 1995. *Non-Places: Introduction to an Anthropology of Supermodernity*, trans. J. Howe. London and New York: Verso.

———. 1998. *A Sense for the Other: The Timeliness and Relevance of Anthropology*, trans. A. Jacobs. Mestizo Spaces. Stanford: Stanford University Press.

———. 1999. *An Anthropology for Contemporaneous Worlds*, trans. A. Jacobs. Mestizo Spaces. Stanford: Stanford University Press.

Augé, M., and J.-P. Colleyn. 2006. *The World of the Anthropologist*, trans. J. Howe. Oxford and New York: Berg.

Austin, J.L. 1962. *How to Do Things with Words*. The William James Lectures Series, 1955. Oxford: Clarendon Press.

Bachelard, G. 1994. *The Poetics of Space: The Classic Look at How We Experience Intimate Spaces*, trans. M. Jolas. Boston, MA: Beacon Press.

Bachnik, J.M. 1998. "Time, Space and Person in Japanese Relationships." In *Interpreting Japanese Society: Anthropological Approaches*, ed. J. Hendry, 2nd ed., pp. 91–116. London and New York: Routledge.

Bachnik, J.M., and J. Quinn, eds. 1994. *Situated Meaning: Inside and Outside in Japanese Self, Society and Language*. Princeton: Princeton University Press.

Bakhtin, M.M. 1981. "Discourse in the Novel." In *The Dialogic Imagination: Four Essays*, ed. M. Holquist, trans. C. Emerson and M. Holquist, pp. 259–422. University of Texas Press Slavic Series, No. 1. Austin: University of Texas Press.

Barthes, R. 1982. *Empire of Signs*, trans. R. Howard. New York: Hill and Wang, A Division of Farrar, Straus and Giroux.

———. 2005. *The Neutral: Lecture Course at the at the* [sic] *Collège de France (1977–1978)*, trans. R.E. Krauss and D. Hollier. New York: Columbia University Press.

Bataille, G. 1989 [1967]. *The Accursed Share: An Essay on General Economy, Vol. 1, Consumption*. New York: Zone Books.

Baudrillard, J. 1998. *The Consumer Society: Myths and Structures*. London: Sage.

Bauman, Z. 2003. *Liquid Love: On the Frailty of Human Bonds*. Cambridge: Polity.

Beck, J., and M. Beck. 1994. *The Change of a Life-Time: Employment Patterns Among Japan's Managerial Elite*. Hawaii: University of Hawaii Press.

Befu, H. 1980. "Alternative Models: The Next Step." *Social Analysis*, Special Issue, Japanese Society: New Appraisals and New Directions, No. 5/6, December: 29–43.

———. 1989. "Four Models of Japanese Society and Their Relevance to Conflict as Applied to Japan." In *Constructs for Understanding Japan*, ed. Y. Sugimoto and R.E. Mouer, pp. 39–66. London and New York: Kegan Paul International.

Behar, R. 1996. *The Vulnerable Observer: Anthropology That Breaks Your Heart*. Boston: Beacon Press.

Ben-Ari, E. 1994. "The Expansion of Japanese Business and Images of the World Order: The Case of Japanese Executives in Singapore." In *Kyoto Conference of Japanese Studies III*, pp. 278–88. Tokyo: International Japanese Culture Research Centre.

Benedikt, M. 1991. "Cyberspace: Some Proposals." In *Cyberspace: First Steps*, ed. M. Benedikt, pp. 119–224. Cambridge, MA, and London: MIT Press.

Bennett, J.W., and I. Ishino. 1963. *Paternalism in the Japanese Economy: Anthropological Studies of Oyabun-Kobun Patterns*. Minneapolis: University of Minnesota Press.

Bernstein, G.L. 1991. "Introduction." In *Recreating Japanese Women 1600–1945*, ed. G.L. Bernstein, pp. 3–30. Berkeley, LA, London: University of California Press.

Berque, A. 1997. *Japan: Cities and Social Bonds*, trans. C. Turner. Yelvertoft Manor, Northants: Pilkington Press.

Bestor, T.C. 1989. *Neighbourhood Tokyo*. Studies of the East Asian Institute Series. Stanford: Stanford University Press.

Bhabha, H.K. 1994. *The Location of Culture*. London and New York: Routledge.

Bingham, M., and S. Gross. 1987. *Women in Japan: From Ancient Times to the Present*. St. Louis Park: Glenhurst.

Bourdieu, P. 1977. *Outline of a Theory of Practice,* trans. R. Nice. Cambridge: Cambridge University Press.

———. 1984. *Distinction: A Social Critique of the Judgement of Taste*, trans. R. Nice. London and New York: Routledge.

———. 1990. *The Logic of Practice*, trans. R. Nice. Cambridge: Polity.

———. 1991. *Language and Symbolic Power*. Cambridge: Polity.

Brinton, M.C. 1992. "Christmas Cakes and Wedding Cakes: The Social Organization of Japanese Women's Life Course." In *Japanese Social Organization*, ed. T.S. Lebra, pp. 79–107. Honolulu: University of Hawaii Press.

———. 1993. *Women and the Japanese Economic Miracle: Gender and Work in Postwar Japan*. Berkeley, LA, London: University of California Press.

———. 2000. "Social Capital in the Japanese Youth Labour Market: Labour Market Policy, Schools and Norms." *Policy Sciences* 33 (3–4): 289–306.

Brown, M.F. 1996. "Resisting Resistance." *American Anthropologist* 98 (4): 729–49.

Butler, J. 1999. "Performativity's Social Magic." In *Bourdieu: A Critical Reader*, ed. R. Shusterman, pp. 113–28. Oxford: Blackwell.

———. 2004. *Undoing Gender*. New York and London: Routledge.

———. 2007. *Gender Trouble: Feminism and the Subversion of Identity*. London and New York: Routledge.

Calhoun, C., E. LiPuma, and M. Postone, eds. 1993. *Bourdieu: Critical Perspectives*. Chicago: University of Chicago Press.

Carter, R., and L. Dilatush. 1976. "Office Ladies." In *Women in Changing Japan*, ed. J. Lebra, R. Paulson, and E. Powers, pp. 75–87. Boulder, CO, Tonbridge, UK: Westview Press.

Chodorow, N.J. 1999. *The Power of Feelings: Personal Meaning in Psychoanalysis, Gender, and Culture*. New Haven and London: Yale University Press.

Chaimowicz, M.C. 1985. *Café du Réve*. Paris: Thames and Hudson.

Clark, R. 1979. *The Japanese Company*. New Haven: Yale University Press.

Clifford, J. 1997. "Spatial Practices: Fieldwork, Travel, and the Disciplining of Anthropology." In *Routes: Travel and Translation in the Late Twentieth Century*, pp. 52–91. Cambridge, MA, London: Harvard University Press.

Clifford, J. and G.E. Marcus eds. 1986. *Writing Culture: The Poetics and Politics of Ethnography*. Berkeley, LA, London: University of California Press.

Coates, K. 2000. "Back in the Race: Japan and the Internet." In *Japan after the Economic Miracle: In Search of New Directions*, ed. P. Bowles and L.T. Woods, Social Indicators Research Series, Vol. 3, pp. 71–84. Dordrecht, Boston, and London: Kluwer.

Cockburn, C. 1991. *In the Way of Women: Men's Resistance to Sex Equality in Organizations.* Basingstoke: Macmillan.

Cohen, A.P. 1985. *The Symbolic Construction of Community.* Key Ideas Series. London and New York: Routledge.

———. 1992. "Self-conscious Anthropology." In *Anthropology and Autobiography,* ed. J. Okely and H. Callaway, pp. 221–40. ASA Monographs 29. London: Routledge.

Cole, R.E. 1971. *Japanese Blue Collar: The Changing Tradition.* Berkeley, LA, London: University of California Press.

———. 1979. *Work, Mobility, and Participation: A Comparative Study of American and Japanese Industry.* Berkeley, LA, London: University of California Press.

Comaroff, J., and J. Comaroff. 1991. *Of Revelation and Revolution: Christianity, Colonialism, and Consciousness in South Africa.* Vol. 1. Chicago and London: University of Chicago Press.

Connell, R.W., and J.W. Messerschmidt. 2005. Hegemonic Masculinity: Rethinking the Concept. *Gender and Society* 19 (6): 829–59.

Cook, A.H., and H. Hayashi. 1980. *Working Women in Japan: Discrimination and Resistance, and Reform.* Cornell International Industry and Labor Relations Report, No. 10. Ithaca: Cornell University Press.

Cornwall, A., and N. Lindisfarne. 1994. *Dislocating Masculinities: Comparative Ethnographies.* New York and London: Routledge.

Crapanzano, V. 1986. "Hermes' Dilemma: The Masking of Subversion in Ethnographic Description." In *Writing Culture: The Poetics and Politics of Ethnography,* ed. J. Clifford and G.E. Marcus, pp. 51–76. Berkeley, LA, London: University of California Press.

———. 1995. "The Postmodern Crisis: Discourse, Parody, Memory." In *Bakhtin in Contexts: Across the Disciplines,* ed. Mandelker, A. and C. Emerson, pp. 137–50. Illinois: Northwestern University Press.

Crowley, T. 2001 [1989]. "Bakhtin and the History of the Language." In *Bakhtin and Cultural Theory,* 2nd ed., ed. Hirschkop, K. and D. Shepherd, pp. 177–200. Manchester and New York: Manchester University Press.

Cummings, W.K. 1980. *Education and Equality in Japan.* Princeton: Princeton University Press.

Dalby, L. 2000 [1983]. *Geisha.* London: Vintage.

Dasgupta, R. 2000. "Performing Masculinities? The 'Salaryman' at Work and Play." *Japanese Studies* 20 (2): 189–200.

Davies, C.A. 2008 [1999]. *Reflexive Ethnography: A Guide to Studying Ourselves and Others.* London and New York: Routledge.

de Certeau, M. 1984. *The Practice of Everyday Life,* trans. S. Rendall. Berkeley, LA, London: University of California Press.

Deleuze, G. 1988. *Foucault,* trans. and ed. S. Hand. London: Athlone Press.

———. 1995 [1990]. *Negotiations 1972–1990,* trans. M. Joughin. European Perspectives Series. New York: Columbia University Press.

Derrida, J. 1990. "Sending: On Representation." In *Transforming the Hermeneutic Context: From Nietzsche to Nancy,* ed. G,L. Ormiston and A.D Schrift, pp. 107–38. New York: SUNY Press.

Doi, T. 1985. *The Anatomy of Self: The Individual Versus Society,* trans. M.A. Harbison. Tokyo, New York, and London: Kodansha International.

Dore, R.P. 1973. *British Factory Japanese Factory: The Origin of National Diversity in Industrial Relations*. Berkeley, LA, Oxford: University of California Press.

————. 1976. *The Diploma Disease: Education, Qualification and Development*. London: Allen and Unwin.

————. 1987. *Taking Japan Seriously: A Confucian Perspective on Leading Economic Issues*. Stanford: Stanford University Press.

Douglas, M. 1996 [1970]. *Natural Symbols: Explorations in Cosmology*. London and New York: Routledge.

————. 1986. *How Institutions Think*. London: Routledge and Kegan Paul.

Dower, J. 1986. *War without Mercy: Race and Power in the Pacific*. London: Faber and Faber.

————. 1993. *Japan in War and Peace: Essays on History, Culture and Race*. London: Harper Collins.

Drentea, P. 1988. "Consequences of Women's Formal and Informal Job Search Methods for Employment in Female-Dominated Jobs." *Gender and Society* 12 (3): 321–38.

Dumont, L. 1980. *Homo Hierarchicus: The Caste System and Its Implications*, trans. M. Sainsbury, L. Dumont, and B. Gulati. Revised English edition. Chicago and London: Chicago University Press.

Duncan, N. 1996. "Renegotiating Gender and Sexuality in Public and Private Spaces." In *Body Space: Destabilizing Geographies of Gender and Sexuality*, ed. N. Duncan, pp. 127–45. London and New York: Routledge.

Durkheim, E. 2001 [1912]. *The Elementary Forms of Religious Life*, trans. C. Cosman. Oxford and New York: Oxford University Press.

Durkheim, E. and M. Mauss. 1969 [1903]. *Primitive Classification*, trans. and ed. R. Needham. London: Routledge.

Eades, J.S., Goodman, R., and Y. Hada. 2005. *The "Big Bang" in Japanese Higher Education: The 2004 Reforms and the Dynamics of Change*. Melbourne: Trans Pacific Press; Beppu, Japan: Ritsumeikan Centre for Asia Pacific Studies, Ritsumeikan Asia Pacific University.

Eagleton, T. 1991. *Ideology: An Introduction*. London and New York: Verso.

————. 2007. "I Contain Multitudes." *London Review of Books* 29 (12) (21 June). Online lrb.co.uk, accessed 23 June 2007.

Eisenstadt, S.N., and E. Ben-Ari, eds. 1990. *Japanese Models of Conflict Resolution*. London and New York: Kegan Paul International.

Emerson, R.M., R.I. Fretz, and L.L. Shaw. 1995. *Writing Ethnographic Fieldnotes*. Chicago and London: University of Chicago Press.

Fernandez-Kelly, M.P., and J. Nash, eds. 1983. *Women, Men and the International Division of Labor*. Albany: State University of New York Press.

Foster, G.M. 1972. "The Anatomy of Envy: A Study in Symbolic Behavior." *Current Anthropology* 13 (2): 165–202.

Foucault, M. 1972. *The Archaeology of Knowledge*, trans. A.M. Sheridan Smith. London: Routledge.

————. 1977. *Discipline and Punish: The Birth of the Prison*, trans. A.M. Sheridan. London: Penguin Books.

————. 1980. *Power/Knowledge: Selected Writings and Other Writings 1972–1977*, trans. C. Gordon, L. Marshall, J. Mepham, and K. Soper, ed. C. Gordon. Brighton, Sussex: Harvester Press.

Foucault, M. 1997 [1982]. "The Subject and Power." In *Reading Gender: Postmodernism, Body, and Marginality*, ed. Y. Uchiyamada, pp. 11–32. Tokyo: International Development Research Institute Foundation for Advanced Studies on International Development.

Fujita, M. 1989. "It's All the Mother's Fault: Childcare and the Socialization of Working Mothers in Japan." *Journal of Japanese Studies* 15 (1): 67–91.

Geertz, C. 1973a. "Ideology as a Cultural System." In *The Interpretation of Cultures: Selected Essays*, pp. 193–233. London: Fontana Press.

———. 1973b. "Thick Description: Toward an Interpretive Theory of Culture." In *The Interpretation of Cultures: Selected Essays*, pp. 3–30. London: Fontana Press.

Gerschick, T.J., and A.S. Miller. 1994. "Gender Identities at the Crossroads of Masculinity and Physical Disability." *Masculinites* 2 (1): 34–55.

Giddens, A. 1984. *The Constitution of Society: An Outline of the Theory of Structuration.* Cambridge: Polity.

———. 1990. *The Consequences of Modernity.* Cambridge: Polity.

———. 1994. *Beyond Left and Right: The Future of Radical Politics.* Cambridge: Polity.

Gill, T. 2003. "When Pillars Evaporate: Structuring Masculinity on the Japanese Margins." In *Men and Masculinities in Contemporary Japan: Dislocating the Salaryman Doxa*, ed. Roberson, J.E. and N. Suzuki, pp. 144–61. London and New York: Routledge.

Gilligan, C. 2003 [1982]. *In a Different Voice: Psychological Theory and Women's Development.* Cambridge, MA, London: Harvard University Press.

Glazener, N. 2001 [1989]. "Dialogic Subversion: Bakhtin, the Novel and Gertrude Stein." In *Bakhtin and Cultural Theory*, 2nd ed., ed. Hirschkop, K. and D. Shepherd, pp. 155–176. Manchester and New York: Manchester University Press.

Gluckman, M. 1963. "Gossip and Scandal." *Current Anthropology* 4 (3): 307–16.

Goffman, E. 1959. *The Presentation of Self in Everyday Life.* Middlesex, UK: Penguin Books.

———. 1963. *Behavior in Public Places: Notes on the Social Organization of Gatherings.* New York: Free Press.

———. 1989. "On Fieldwork." *Journal of Contemporary Ethnography* 18: 123–32.

Golden, A. 1997. *Memoirs of a Geisha.* London: Vintage.

Goodman, R. 1990. *Japan's "International Youth": The Emergence of a New Class of School Children.* Oxford: Clarendon Press.

Goodman, R., and K. Refsing. 1992. *Ideology and Practice in Modern Japan.* London and New York: Routledge.

Gordon, A. 1985. *The Evolution of Labor Relations in Japan: Heavy Industry 1853–1955.* Cambridge, MA: Harvard University Press.

Gottfried, H. 2003. "Temp(t)ing Bodies: Shaping Gender at Work in Japan." *Sociology* 37 (2): 257–76.

Gottfried, H., and N. Hayashi-Kato. 1998. "Gendering Work: Deconstructing the Narrative of the Japanese Economic Miracle." *Work, Employment and Society* 12 (1): 25–46.

Granovetter, M. 1974. *Getting a Job*, 2nd ed., 1995. Chicago: University of Chicago Press.

Greer, G. 1991 [1970]. *The Female Eunuch.* London: Flamingo, Harper Collins.

Grint, K. 1991. *The Sociology of Work: An Introduction.* Cambridge: Polity.

Grosz, E. 1994. *Volatile Bodies: Toward a Corporeal Feminism*. Bloomington and Indianapolis: Indiana University Press.

Gullestad, M. 1984. *Kitchen-Table Society*. New York: Columbia University Press.

Gupta, A., and J. Ferguson. 1992. "Beyond 'Culture': Space, Identity, and the Politics of Difference." *Cultural Anthropology* 7 (1): 6–23.

Halford, S., and S. Leonard. 2001. *Gender, Power and Organizations: An Introduction*. London: Palgrave.

Hamabata, M.M. 1990. *Crested Kimono: Power and Love in the Japanese Business Family*. Ithaca and London: Cornell University Press.

Hamada, T. 1985. "Corporation, Culture, and Environment: The Japanese Model." *Asian Survey* 25 (12): 1214–28.

———. 1992. "Under the Silk Banner: The Japanese Company and Its Overseas Managers." In *Japanese Social Organization*, ed. T.S. Lebra, pp. 135–64. Honolulu: University of Hawaii Press.

———. 1996. "Absent Fathers, Feminized Sons, Selfish Mothers and Disobedient Daughters: Revisiting the Japanese Ie Household." *Japan Policy Research Institute Working Paper*, No. 33, accessed online www.jpri.org.

Hamaguchi, E. 1985. "A Contextual Model of the Japanese: Toward a Methodological Innovation in Japanese Studies," trans. S. Kumon and M.R. Creighton. *Journal of Japanese Studies* 11 (2): 289–321.

Hammersley, M., and P. Atkinson. 1995 [1983]. *Ethnography: Principles in Practice*, 2nd ed. London and New York: Routledge.

Hanks, W.F. 1989. "Text and Textuality." *Annual Review of Anthropology* 18: 95–127.

Hardt, M., and A. Negri. 2000. *Empire*. Cambridge, MA, London: Harvard University Press.

Harvey, D. 1989. *The Condition of Postmodernity: An Enquiry into the Origins of Cultural Change*. Cambridge, MA, Oxford, UK: Blackwell.

Heidegger, M. 1962. *Being and Time*, trans. Macquarrie, J. and E. Robinson. Malden MA, Oxford, UK, and Victoria, Australia: Blackwell.

Hendry, J. 1986. *Becoming Japanese: The World of the Pre-school Child*. Manchester: Manchester University Press.

———. 1990. "Humidity, Hygiene, or Ritual Care: Some Thoughts on Wrapping as a Social Phenomenon." In *Unwrapping Japan: Society and Culture in Anthropological Perspective*, ed. Ben-Ari, E., B. Moeran, and J. Valentine, pp. 18–35. Manchester: Manchester University Press.

———. 1992. "The Paradox of Friendship in the Field: Analysis of a Long-Term Anglo-Japanese Relationship." In *Anthropology and Autobiography*, ed. Okely, J. and H. Callaway, pp. 163–74. ASA Monographs 29. London and New York: Routledge.

———. 1993a. "The Role of the Professional Housewife." In *Japanese Women Working*, ed. J. Hunter, pp. 224–41. London and New York: Routledge.

———. 1993b. *Wrapping Culture: Politeness, Presentation, and Power in Japan and Other Societies*. Oxford: Clarendon.

———. 1995 [1987]. *Understanding Japanese Society*, 2nd ed. London and New York: Routledge.

Hettinger, J.F., and S.D. Tooley. 1994. *Small Town, Giant Corporation: Japanese Manufacturing Investment and Community Economic Development in the United States.* Lanham, MD: University Press of America.

Hicks, H.G., and C.R. Gullet. 1975. *Organizations: Theory and Behavior.* New York: McGraw-Hill.

Highmore, B. 2002. "Introduction: Questioning Everyday Life." In *The Everyday Life Reader*, ed. B. Highmore, pp. 1–34. London and New York: Routledge.

Hill, J. 1986. "Review: The Refiguration of the Anthropology of Language." *Cultural Anthropology* 1 (1): 89–102.

Hine, C. 2000. *Virtual Ethnography.* London, Thousand Oaks, and New Delhi: Sage.

Hirsch, E., and Gellner, D.N. 2001. "Introduction: Ethnography of Organizations and Organizations of Ethnography." In *Inside Organizations: Anthropologists at Work*, ed. D.N. Gellner and E. Hirsch, pp. 1–15. Oxford and New York: Berg.

Hite, S. 2000. *Sex and Business: Ethic of Sexuality in Business and the Workplace.* Financial Times/Prentice Hall.

Hochschild, A.R. 1983. *The Managed Heart: Commercialization of Human Feeling.* Berkeley, LA: University of California Press.

Holzhausen, A. 2000. "Japanese Employment Practices in Transition: Promotion Policy and Compensation Systems in the 1990s." *Social Science Japan Journal* 3 (2): 221–35.

Hunter, J. 1993. "Introduction." In *Japanese Women Working*, ed. J. Hunter, pp. 1–15. London and New York: Routledge.

Imamura, A.E. 1987. *Urban Japanese Housewives: At Home and in the Community.* Honolulu: University of Hawaii Press.

Ingold, T. 1999. Paper presented at anthropology departmental seminar, SOAS, University of London, October, n.p.

Ishida, H. 1993. *Social Mobility in Contemporary Japan: Education Credentials, Class, and the Labor Market in a Cross-national Perspective.* Stanford: Stanford University Press.

Ishii-Kuntz, M. 2003. "Balancing Fatherhood and Work: Emergence of Diverse Masculinities in Contemporary Japan." In *Men and Masculinities in Contemporary Japan: Dislocating the Salaryman Doxa*, ed. Roberson, J.E. and N. Suzuki, pp. 198–216. London and New York: Routledge.

Ivy, M. 1995. *Discourses of the Vanishing: Modernity, Phantasm, Japan.* Chicago and London: University of Chicago Press.

Iwai, T. 1993. "'The Madonna Boom': Women in the Japanese Diet." *Journal of Japanese Studies* 19 (1): 103–20.

Iwao, S. 1993. *The Japanese Woman: Traditional Image and Changing Reality.* Cambridge, MA: Harvard University Press.

Jacoby, S.M. 2007. *The Embedded Corporation.* Princeton: Princeton University Press.

Johnson, C. 1982. *MITI and the Japanese Miracle: The Growth of Industrial Policy, 1925–1975.* Stanford: Stanford University Press.

Jordan, T. 1999. *Cyberpower: The Culture and Politics of Cyberspace and the Internet.* London and New York: Routledge.

Kanter, R.M. 1977. *Men and Women of the Corporation.* New York: Basic Books.

Kapferer, B. 1988. "The Anthropologist as Hero: Three Exponents of Post-modernist Anthropology." Review article. *Critique of Anthropology* 8 (2): 77–104.

Kelsky, K. 2001. *Women on the Verge: Japanese Women, Western Dreams*. Durham and London: Duke University Press.

Kilian, T. 1998. "Public and Private, Power and Space." In *Philosophy and Geography II: The Production of Public Space*, ed. A. Light and J.M. Smith, pp. 115–33. Lanham, Boulder, New York, and Oxford: Rowman and Littlefield.

Kingston, J. 2004. *Japan's Quiet Transformation: Social Change and Civil Society in the Twenty-First Century*. London: RoutledgeCurzon.

Kinzley, W.D. 1991. *Industrial Harmony in Modern Japan: The Invention of a Tradition*. Nissan Institute/Routledge Japanese Studies Series. London and New York: Routledge.

Kitahara, M. 1994. "Honorific Ambiguity and Conflict in Japan." *Sociologus* 44 (2): 179–189.

Klein, R. 1993. *Cigarettes Are Sublime*. Durham and London: Duke University Press.

Komashaku, K. ed. 1985. 女を装う (Simulating Woman). Tokyo: Komakusa Shobo.

Kondo, D.K. 1986. "Dissolution and Reconstruction of Self: Implications for Anthropological Epistemology." *Cultural Anthropology* (1): 74–88.

———. 1990a. *Crafting Selves: Power, Gender, and Discourses of Identity in a Japanese Workplace*. Chicago and London: University of Chicago Press.

———. 1990b. "M. Butterfly: Orientalism, Gender, and a Critique of Essentialist Identity." *Cultural Critique* (16), Fall: 5–30.

Konno, M. 2000. *OL* の創造：意味世界としてのジェンダー (The Construction of OLs: Gender as a World of Meaning). Tokyo: Keiso Shobo.

Krauss, E.S., T.P. Rohlen, and P.G. Steinhoff, eds. 1984. *Conflict in Japan*. Honolulu: University of Hawaii Press.

Kristeva, J. 1980. "Word, Dialogue, and Novel." In *Desire in Language: A Semiotic Approach to Literature and Art*, ed. L.S. Roudiez, trans. T. Gora, A. Jardine, and L.S. Roudiez, pp. 64–91. New York: Columbia University Press.

———. 1991. *Strangers to Ourselves*, trans. L.S. Roudiez. New York: Columbia University Press.

Kurihara, T. 2007. "Seken." In The Blackwell Encyclopaedia of Sociology, ed. G. Ritzer, Vol. VIII. pp. 4154–7. Oxford: Blackwell.

Kuwayama, T. 1992. "The Reference Other Orientation." In *Japanese Sense of Self*, ed. N. Rosenberger, pp. 121–51. Cambridge: Cambridge University Press.

Lam, A. 1992. *Women and Japanese Management: Discrimination and Reform*. London and New York: Routledge.

Latour, B. 1996. *Aramis or the Love of Technology*, trans. C. Porter. Cambridge, MA, London: Harvard University Press.

Latour, B., and S. Woolgar. 1986. *Laboratory Life: The Construction of Scientific Facts*. Chichester and New Jersey: Princeton University Press.

Layton, R. 1997. *An Introduction to Theory in Anthropology*. Cambridge: Cambridge University Press.

Lebra, J., R. Paulson, and E. Powers. 1976. *Women in Changing Japan*. Boulder: Westview Press.

Lebra, T.S. 1976. *Japanese Patterns of Behavior*. Honolulu: University of Hawaii Press.

———. 1981. "Japanese Women in Male Dominant Careers: Cultural Barriers and Accommodations for Sex-Role Transcendence." *Ethnology* (20): 291–306.

———. 1984. *Japanese Women: Constraint and Fulfilment*. Honolulu: University of Hawaii Press.

Lebra, T.S. 1992a. "The Spatial Layout of Hierarchy: Residential Style of the Modern Nobility." In *Japanese Social Organization*, ed. T.S. Lebra, pp. 49–78. Honolulu: University of Hawaii Press.

———. 1992b. "Self in Japanese Culture." In *Japanese Sense of Self*, ed. N. Rosenberger, pp. 105–20. Cambridge: Cambridge University Press.

———. 1993. *Above the Clouds: Status Culture of the Modern Japanese Nobility*. Berkeley, LA, London: University of California Press.

Lefebvre, H. 1991 [1974]. *The Production of Space*, trans. D. Nicholson-Smith. Oxford and London: Blackwell.

Levi-Strauss, C. 1963. *Structural Anthropology*. New York: Basic Books.

———. 1973. *Tristes Tropiques*, trans. J and D. Weightman. London: Penguin.

Lin, N., K. Cook, and R.S. Burt eds. 2001. *Social Capital: Theory and Research*. New Brunswick and London: Aldine Transaction, A Division of Transaction Publishers.

Lo, J. 1990. *Office Ladies, Factory Women: Life and Work at a Japanese Company*. Armonk, New York, and London: M.E. Sharpe.

Lock, M. 1988. "New Japanese Mythologies: Faltering Discipline and the Ailing Housewife." *American Ethnologist* 15 (1): 43–61.

Long, N. 1996. "Globalization and Localization: New Challenges in Rural Research." In *The Future of Anthropological Knowledge*, ed. H.L. Moore, pp. 37–59. ASA Decennial Conference Series. The Uses of Knowledge: Global and Local Relations. London and New York: Routledge.

Low, M., S. Nakayama, and H. Yoshida. 1999. *Science, Technology and Society in Contemporary Japan*. Contemporary Japanese Society Series. Cambridge: Cambridge University Press.

Lowe, L. 1991. *Critical Terrains: French and British Orientalisms*. Ithaca: Cornell University Press.

Mackie, V. 1995. "Equal Opportunity and Gender Identity: Feminist Encounters with Modernity and Postmodernity in Japan." In *Japanese Encounters with Postmodernity*, ed. Y. Sugitomo and J.P. Arnason. London: Kegan Paul International.

Marcus, G.E. 1995. "Ethnography in/of the World System: The Emergence of Muti-sited Ethnography." *Annual Review of Anthropology* 24: 95–117.

———. 1999. "Critical Anthropology Now: An Introduction." In *Critical Anthropology Now: Unexpected Contexts, Shifting Constituencies, Changing Agendas*, ed. G.E. Marcus, pp. 3–28. Santa Fe, NM: School of American Research Press.

Margolis, J. 1999. "Pierre Bourdieu: Habitus and the Logic of Practice." In *Bourdieu: A Critical Reader*, ed. R. Shusterman, pp. 64–83. Malden, MA, Oxford: Blackwell.

Martinez, D.P. 1993. "Women as Bosses: Perceptions of the *Ama* and Their Work." In *Japanese Women Working*, ed. J. Hunter, pp. 181–96. London and New York: Routledge.

———. 1998. "Introduction: Gender, Shifting Boundaries and Global Cultures." In *The Worlds of Japanese Popular Culture: Gender, Shifting Boundaries and Global Cultures*, ed. D.P. Martinez, pp. 1–18. Cambridge: Cambridge University Press.

Marx, K. 1967 [1867]. *Capital*. London: Lawrence and Wishart.

Mathews, G. 2003. "Can 'a Real Man' Live for His Family?: *Ikigai* and Masculinity in Today's Japan." In *Men and Masculinities in Contemporary Japan: Dislocating the Salaryman Doxa*, ed. J.E. Roberson and N. Suzuki, pp. 109–25. Nissan Institute/RoutledgeCurzon Japanese Studies Series. London and New York: Routledge.

Matsunaga, L. 1995. *Working in a Japanese Chain Store: An Ethnography of a Japanese Company*. Ph. D. diss. University of London.

———. 2000. *The Changing Face of Japanese Retail: Working in a Chain Store*. Nissan Institute/Routledge Japanese Studies Series. London and New York: Routledge.

Mauss, M. 1979 [1935]. "The Notion of Body Techniques." In *Sociology and Psychology: Essays by Marcel Mauss*, trans. B. Brewster, pp. 95–123. London: Routledge and Kegan Paul.

———.1990 [1925]. *The Gift: The Form and Reason for Exchange in Archaic Societies*, trans. W.D. Halls. London and New York: Routledge.

McDowell, L. 1997. *Capital Culture: Gender at Work in the City*. Oxford: Blackwell.

McKinstry, J.A., and A.N. McKinstry. 1991. *Jinsei Annai "Life's Guide": Glimpses of Japan through a Popular Advice Column*. New York and London: M.E. Sharpe.

McLendon, J. 1983. "The Office: Way Station or Blind Alley?" In *Work and Life Course in Japan*, ed. D.W. Plath. Albany: State University of New York.

McNay, L. 2004. "Agency and Experience: Gender as a Lived Relation." In *Feminism After Bourdieu*, ed. L. Adkins and B. Skeggs, pp. 175–90. Oxford, UK, Malden, MA: Blackwell Sociological Review.

McVeigh, B.J. 1997. *Life in a Japanese Women's College: Learning to Be Ladylike*. Nissan Institute/Routledge Japanese Studies Series. London and New York: Routledge.

Meillassoux, C. 1981 [1975]. *Maidens, Meal and Money: Capitalism and the Domestic Community*. Cambridge: Cambridge University Press.

Merleau-Ponty, M. 2005 [1958]. *Phenomenology of Perception*, trans. C. Smith. London and New York: Routledge.

Miller, L. 2003. "Male Beauty Work in Japan." In *Men and Masculinities in Contemporary Japan: Dislocating the Salaryman Doxa*, ed. J.E. Roberson and N. Suzuki, pp. 37–58. London and New York: Routledge.

Minear, R. 1980. "Orientalism and the Study of Japan." *Journal of Asian Studies* 39 (3): 507–17.

Mintz, S.W., and C.M. Du Bois. 2002. "The Anthropology of Food and Eating." *Annual Review of Anthropology* 31: 99–119.

Mitchell, J. 2003. *Siblings: Sex and Violence*. Cambridge: Polity.

Miura, A. 2005. 下流社会：新たな階層集団の出現 (Karyū-shakai: The Appearance of a New Social Stratum). Tokyo: Kobunsha.

Miyoshi, M. 1991. *Off Center: Power and Cultural Relations between Japan and the United States*. Cambridge, MA: Harvard University Press.

———. ed. 1993. *Japan in the World*. Durham, NC, London: Duke University Press.

Miyoshi, M., and H.D. Harootunian, eds. 1989. *Post-modernism and Japan*. Durham, NC, London: Duke University Press.

Moeran , B. 1990. "Introduction: Rapt Discourses: Anthropology, Japanism and Japan." In *Unwrapping Japan: Society and Culture in Anthropological Perspective*, ed. E. Ben-Ari, B. Moeran, and J. Valentine, pp. 1–17. Manchester: Manchester University Press.

Moi, T. 2005. *Sex, Gender and the Body: The Student Edition of What Is a Woman?* Oxford and New York: Oxford University Press.

Moore, H.L. 1986. *Space, Text, and Gender: An Anthropological Study of the Marakwet of Kenya*. Cambridge: Cambridge University Press.

———. 1994. *A Passion for Difference: Essays in Anthropology and Gender*. Cambridge: Polity.

Moore, H.L. 1996. "The Changing Nature of Anthropological Knowledge: An Introduction." In *The Future of Anthropological Knowledge*, ed. H.L. Moore, pp. 1–15. ASA Decennial Conference Series. London and New York: Routledge.

Moore, H., and T. Sanders. 2006. "Anthropology and Epistemology." In *Anthropology in Theory: Issues in Epistemology*, ed. H.L. Moore and T. Sanders, pp. 1–21. Oxford: Blackwell.

Morinaga, E. 1995. "企業中心社会の抑圧： 企業という '車座社会'のなかで" ("Oppressive Corporate Centered Society: Amidst an 'Insular Society' of the Corporation)." In *Feminism in Japan: Studies in Masculinity*, Separate Vol., ed. T. Inoue, C. Ueno, and Y. Ehara, pp. 217–20. Tokyo: Iwanami Shoten.

Mouer, R.E., and H. Kawanishi. 2005. *A Sociology of Work in Japan*. Cambridge and New York: Cambridge University Press.

Mouer, R. E., and Y. Sugimoto. 1980. "Competing Models for Understanding Japanese Society: Some Reflections on New Directions." *Social Analysis*, Special Issue, Japanese Society: Reappraisals and New Directions, No. 5/6, December: 194–204.

———. eds. 1989. *Constructs for Understanding Japan*. London and New York: Kegan Paul International.

Munn, N.D. 1992. "The Cultural Anthropology of Time: A Critical Essay." *Annual Review of Anthropology* 21: 93–123.

Nadar, L. 1969. "Up the Anthropologist—Perspectives Gained from Studying Up." In *Reinventing Anthropology*, ed. D. Hymes, pp. 284–311. New York: Pantheon, Random House.

Nakamura, K. 2006. *Deaf in Japan: Signing and the Politics of Identity*. Ithaca and London: Cornell University Press.

Nakane, C. 1970. *Japanese Society*. Tokyo: Charles E. Tuttle.

Nancy, J.-L. 1991. *The Inoperative Community*, ed. P. Connor, trans. P. Connor, L. Garbus, M. Holland, and S. Sawhney. Theory and History of Literature, Vol. 76. Minneapolis and London: University of Minnesota Press.

Napier, S.J. 1991. *Escape from the Wasteland: Romanticism and Realism in the Fiction of Mishima Yukio and Oe Kenzaburo*. Cambridge, MA: Harvard University Press.

———. 1998. "Vampires, Psychic Girls, Flying Women and Sailor Scouts: Four Faces of the Young Female in Japanese Popular Culture." In *The Worlds of Popular Culture: Gender, Shifting Boundaries and Global Cultures*, ed. D.P. Martinez, pp. 91–109. Cambridge: Cambridge University Press.

Nash, K. 1994. "The Feminist Production of Knowledge: Is Deconstruction a Practice for Women?" *Feminist Review* 47: 65–77.

Nathan, J. 2004. *Japan Unbound: A Volatile Nation's Quest for Pride and Purpose*. Boston: Houghton Mifflin.

Noguchi, P.H. 1990. *Delayed Departures, Overdue Arrivals: Industrial Familism and the Japanese National Railways*. Honolulu: University of Hawaii Press.

Nonaka, I., and H. Takeuchi. 1995. *The Knowledge-Creating Company: How Japanese Companies Create the Dynamics of Innovation*. New York and Oxford: Oxford University Press.

Nye, R.A. 2005. "Locating Masculinity: Some Recent Work on Men." *Signs: Journal of Women in Culture and Society* 30 (3): 1937–62.

Oë, K. 1995. *Japan, the Ambiguous, and Myself: The Nobel Prize Speech and Other Lectures*. Tokyo and New York: Kodansha International.

Ogasawara, Y. 1998. *Office Ladies and Salaried Men: Power, Gender, and Work in Japanese Companies*. Berkeley, LA, London: University of California Press.

Okely, J. 1992. "Anthropology and Autobiography: Participatory Experience and Embodied Knowledge." In *Anthropology and Autobiography*, ed., J. Okely and H. Callaway, pp. 1–28. ASA Monographs 29. London and New York: Routledge.

Ong, A. 1987. *Spirits of Resistance and Capitalist Discipline: Factory Women in Malaysia*. Albany: University of New York Press.

Ortner, S.B. 1984. "Theory and Anthropology Since the Sixties." *Comparative Studies in Society and History* 26 (1): 126–66.

Osawa, H. 2000. ジェンダー関係の日本的構造. (Japanese Structure of Gender Relations). Tokyo: Kobundo Shuppansha.

Osawa, Machiko. 1994. "Japanese-Style Employment Practices and Male-Female Wage Differentials." *Japanese Economic Studies* 22 (5–6): 3–43.

Osawa, Mari. 1993. "Feminization of Employment in Japan." *Annals of the Institute of Social Science* 34 (March): 47–70.

———. 1994. "Bye-Bye Corporate Warriors: The Formation of a Corporate-Centered Society and Gender-Biased Social Policies in Japan." *Annals of the Institute of Social Science* 35 (March): 157–94.

———. 1996. "Will the Japanese Style Employment System Change? Employment, Gender and the Welfare State," trans. R. LeBlanc. *Journal of Pacific Asia* 3: 69–94.

———. 2000. "Government Approaches to Gender Equality in the mid-1990s." *Social Science Japan Journal* 3 (1): 3–19.

Painter, A.A. 1993. "Japanese Daytime Television, Popular Culture, and Ideology." *Journal of Japanese Studies* 19 (2): 295–325.

Parsons, T. 1975. "Some Theoretical Considerations on the Nature and Trends of Change in Ethnicity." In *Ethnicity: Theory and Experience*, ed. N. Glazer and D.P. Moynihan, pp. 53–83. Cambridge, MA: Harvard University Press.

Pascal, R., and A. Athos. 1981. *The Art of Japanese Management*. New York: Warner Books.

Passin, H. 1965. *Society and Education in Japan*. New York: Bureau of Publications, Teachers College: Columbia University.

Peterson, D.E. 1995. "Response and Call: The African American Dialogue with Bakhtin and What It Signifies." In *Bakhtin in Contexts: Across the Disciplines*, ed. Mandelker, A. and C. Emerson, pp. 89–98. Illinois: Northwestern University Press.

Pharr, S.J. 1990. *Losing Face: Status Politics in Japan*. Berkeley, LA, London: University of California Press.

Pierce, J.L. 1995. *Gender Trials: Emotional Lives in Contemporary Law Firms*. Berkeley, LA: University of California Press.

Plath, W.D. 1964. *The After Hours: Modern Japan and the Search for Enjoyment*. Berkeley, LA, London: University of California Press.

———. 1980. *Long Engagements: Maturity in Modern Japan*. Stanford: Stanford University Press.

Poster, M. 1990. *The Mode of Information: Post-structuralism and Social Context*. Cambridge: Polity.

Pringle, R. 1988. *Secretaries Talk: Sexuality, Power and Work*. London: Verso.

———. 1994. "Office Affairs." In *Anthropology of Organizations*, ed. S. Wright, pp. 115–23. London and New York: Routledge.

Reischauer, E.O. 1977. *The Japanese*. Cambridge, MA, London: Harvard University Press.

Ridgeway, C.L., and S.J. Correll. 2004. "Unpacking the Gender System: A Theoretical Perspective on Gender Beliefs and Social Relations." *Gender and Society* 18 (4): 510–31.

Riley, B. 1988. *"Am I That Name?": Feminism and the Category of "Women" in History*. New York and Basingstoke: Palgrave Macmillan.

———. 2005. *Impersonal Passion: Language as Affect*. Durham and London: Duke University Press.

Roberson, J.E. 1995. "Becoming *Shakaijin*: Working-Class Reproduction in Japan." *Ethnology* 34 (4): 293–313.

———. 2003. "Japanese Working-Class Masculinities: Marginalized Complicity." In *Men and Masculinities in Contemporary Japan: Dislocating the Salaryman Doxa*, ed. J.E. Roberson and N. Suzuki, pp. 126–43. London and New York: Routledge.

Roberts, G.S. 1994. *Staying on the Line: Blue-Collar Women in Contemporary Japan*. Honolulu: University of Hawaii Press.

Rohlen, T.P. 1974. *For Harmony and Strength: Japanese White-Collar Organization in Anthropological Perspective*. Berkeley, LA, London: University of California Press.

———. 1980. "The *Juku* Phenomenon: An Explanatory Essay." *Journal of Japanese Studies* 6 (2): 207–42.

———. 1983. *Japan's Highschools*: Berkeley, LA, London: University of California Press.

Rojewski, J.W. 1999. "Vocational Preparation and Employment Options for Adults with Disabilities: An International Perspective." In *Adults with Disabilities: International Perspectives in the Community*, ed. P. Retish and S. Reiter, pp. 229–56. Philadelphia: Laurence Erlbaum Associates.

Rosenberger, N., ed. 1992. *Japanese Sense of Self*. Cambridge: Cambridge University Press.

———. 1995. "Antiphonal Performances? Japanese Women's Magazines and Women's Voices." In *Women, Media and Consumption in Japan*, ed. L. Skov and B. Moeran, pp. 143–69. Surrey: Curzon Press.

Rupp, K. 2003. *Gift-Giving in Japan: Cash, Connections, Cosmologies*. Stanford: Stanford University Press.

Said, E.W. 1995 [1978]. *Orientalism: Western Conceptions of the Orient* (New Afterword). London: Penguin Books.

Saso, M. 1990. *Women in the Japanese Workplace*. London: Hilary Shipman.

Sassen, S. 1990. *The Global City: New York, London and Tokyo*. Princeton: Princeton University Press.

Schaede, U. 1995. "The 'Old Boy' Network and Government-Business Relationships in Japan." *Journal of Japanese Studies* 21 (2): 293–317.

Schindler, P., and T. Cher. 1993. "The Structure of Interpersonal Trust in the Workplace." *Psychological Reports* 73: 563–73.

Scott, J.C. 1985. *Weapons of the Weak: Everyday Forms of Peasant Resistance*. New Haven and London: Yale University Press.

Sedgwick, M.W. 2001. "Positioning 'Globalization' at Overseas Subsidiaries of Japanese Multinational Corporations." In *Globalizing Japan: Ethnography of the Japanese Presence in Asia, Europe, and America*, ed. H. Befu and S. Guichard-Anguis, pp. 43–51. London and New York: Routledge.

———. 2007. *Globalization and Japanese Organizational Culture: An Ethnography of the Japanese Corporation in France*. Japanese Anthropology Workshop Series. London and New York: Routledge.

Seiyama, K. 1994. "Labour Market and Career Mobility." In *Social Stratification in Contemporary Japan*, ed. K. Kosaka, pp. 78–92. London and New York: Kegan Paul.

Sennett, R. 1998. *The Corrosion of Character: The Personal Consequences of Work in the New Capitalism*. New York and London: W.W. Norton.

———. 2008. *The Craftsman*. New Haven and London: Yale University Press.

Shusterman, R., ed. 1999. *Bourdieu: A Critical Reader*. Oxford and Malden, MA: Blackwell.

Sievers, S. 1983. *Flowers in Salt: The Beginnings of Feminist Consciousness in Modern Japan*. Stanford: Stanford University Press.

Simmel, G. 1971 [1908]. *George Simmel on Individuality and Social Forms*, ed. D. Levine. Chicago: Chicago University Press.

Skeggs, B. 1998. *Formations of Class and Gender: Becoming Respectable*. London: Sage.

Sklair, L. 2001. *The Transnational Capitalist Class*. Oxford: Blackwell.

Skov, L., and B. Moeran. 1995. "Introduction: Hiding in the Light: From Oshin to Yoshimoto Banana." In *Women, Media and Consumption in Japan*, ed. L. Skov and B. Moeran, pp. 1–74. ConsumAsiaN Book Series. Surrey: Curzon Press.

Smith, H.D. 1978. "Tokyo as an Idea: An Exploration of Japanese Urban Thought Until 1945." *Journal of Japanese Studies* (4): 44–80.

Smith, R.J. 1983. *Japanese Society: Tradition, Self, and the Social Order*. Lewis Henry Morgan Lecture Series, 1980. New York and Cambridge: Cambridge University Press.

———. 1987. "Gender Inequality in Contemporary Japan." *Journal of Japanese Studies* 13 (1): 1–25.

Soja, E. 1989. *Postmodern Geographies: The Reassertion of Space in Critical Social Theory*. London: Verso.

Spivak, G.C. 1997 [1976]. "Translator's Preface." In *Of Grammatology*, trans. G.C. Spivak, pp. ix–lxxxvii. Baltimore and London: Johns Hopkins University Press.

Stephens, D.L., Collins, M.D., and R.A. Dodder. 2005. "A Longitudinal Study of Employment and Skill Acquisition among Individuals with Developmental Disabilities." *Research in Developmental Disabilities: A Multidisciplinary Journal* 26 (5): 469–86.

Suehiro, A. 1998. "An Introduction to This Issue's Special Topic: Japanese Society and 'Community.'" *Social Science Japan Journal* 1 (2): 163–4.

Sugimoto, Y. 1997. *An Introduction to Japanese Society*. Cambridge, New York, and Melbourne: Cambridge University Press.

Sunderland, P.L. 1999. "Fieldwork and the Phone." *Anthropological Quarterly* 72 (3): 105–17.

Tabata, H. 1998. "Community and Efficiency in the Japanese Firm." *Social Science Japan Journal* 1 (2): 199–215.

Takenobu, M. 1994. 日本株式会社の女たち：総合職たち (Women in Japanese Corporations: The Career Track). In 日本のフェミニズム ４：権力と労働 *(Feminism in Japan: Power and Work)*, Vol. 4, ed. T. Inoue, C. Ueno, and Y. Ehara, pp. 40–9. Tokyo: Iwanami Shoten.

Tanaka, K. 1990. "Intelligent Elegance: Women in Japanese Advertising." In *Unwrapping Japan: Society and Culture in Anthropological Perspective*, ed. E. Ben-Ari, B. Moeran, and J. Valentine, pp. 78–96. Manchester: Manchester University Press.

———. 1994. *Advertising Language: A Pragmatic Approach to Advertisements in Britain and Japan*. London and New York: Routledge.

Tannebaum, L. 2002. *Catfight: Rivalries among Women: From Diets to Dating, From the Boardroom, to the Delivery Room*. New York: Seven Stories Press.

Taylor, C. 1999. "To Follow a Rule..." In *Bourdieu: A Critical Reader*, ed. R. Shusterman, pp. 29–44. Oxford: Blackwell.

Testart, A. 1998. "Uncertainties of the 'Obligation to Reciprocate': A Critique of Mauss." In *Marcel Mauss: A Centenary Tribute*, pp. 97–110. Oxford and New York: Berghahn Books.

Tinkler, I., ed. 1990. *Persistent Inequalities: Women and World Development*. New York and Oxford: Oxford University Press.

Tobin, J.J. 1992. "Japanese Pre-schools and Pedagogy of Selfhood." In *Japanese Sense of Self*, ed. N. Rosenberger, pp. 21–39. Cambridge: Cambridge University Press.

Tönnies, F. 1955 [1887]. *Community and Association (Gemeinschaft and Gesellschaft)*, trans. C.P. Loomis. London: Routledge and Kegan Paul.

Traweek, S. 1988. *Beamtimes and Lifetimes: The World of High Energy Physicists*. Cambridge, MA, London: Harvard University Press.

———. 1995. "Bodies of Evidence: Law and Order, Sexy Machines, and the Erotics of Fieldwork among Physicians." In *Choreographing History*, ed. S.L. Foster. Bloomington Indiana University Press.

Treat, J.W. 1993. "Yoshimoto Banana Writes Home: *Shōjo* Culture and the Nostalgic Subject." *Journal of Japanese Studies* 19 (2): 353–87.

Tsurumi, E.P. 1984. "Female Textile Workers and the Failure of Early Trade Unionism in Japan." *History Workshop* (18): 3–27.

Turner, C.L. 1995. *Japanese Workers in Protest: An Ethnography of Consciousness and Experience*. Berkeley, LA, London: University of California Press.

Turner, V.W. 1969. *The Ritual Process: Structure and Anti-Structure*. The Lewis Henry Morgan Lectures. London and New York: Routledge.

Ueno, C. 1987a. "The Position of Japanese Women Reconsidered." *Current Anthropology* 28 (4): S75–S84.

———. 1987b. "世紀末ウォッチング" (Watching the End of the Century). In 〈私〉探しゲーム：欲望私民社会論 (*A Game in Search of < Self >: A Social Theory of Desire and Citizenship)*, pp. 14–51. Tokyo: Chikuma Shobo.

———. 1995. Untitled Introduction to Section V "企業戦士たち" (Corporate Warriors). In *Feminism in Japan*: 男性学 (*Masculinity Studies)*, Separate Vol., ed. T. Inoue, C. Ueno, and Y. Ehara. p. 216. Tokyo: Iwanami Shoten.

Uno, K. 1993. "One Day at a Time: Work and Domestic Activities of Urban Lower-Class Women in Early Twentieth-Century Japan." In *Japanese Women Working*, ed. J. Hunter, pp. 37–68. London and New York: Routledge.

Valentine, J. 1990. "On the Borderlines: The Significance of Marginality in Japanese Society." In *Unwrapping Japan: Society and Culture in Anthropological Perspective*, ed. E. Ben-Ari, B. Moeran, and J. Valentine, pp. 36–57. Manchester: Manchester University Press.

van Wolferen, K. 1989. *The Enigma of Japanese Power: People and Politics in a Stateless Nation.* London: Macmillan.

Vogel, E.F. 1963. *Japan's New Middle Class: The Salary Man and His Family in a Tokyo Suburb.* 2nd ed. Berkeley, LA, London: University of California Press.

———. ed. 1975. *Modern Japanese Organization and Decision-Making.* Berkeley, LA, London: University of California Press.

———. 1979. *Japan as Number One: Lessons for America.* Cambridge, MA: Harvard University Press.

Vogel, S.H. 1978. "Professional Housewife: The Career of Urban Middle Class Japanese Women." *Japan Interpreter* 12 (1): 16–43.

Warner, W.L. 1941. "Social Anthropology and the Modern Community." *American Journal of Sociology* 46: 785–96.

Watanabe, T. 1995. "抑圧された男性 (Men Under Pressure)." *In Feminism in Japan*: 男性学 (*Masculinity Studies*), separate Vol., ed. T. Inoue, C. Ueno, and Y. Ehara. pp. 109–130. Tokyo: Iwanami Shoten.

Weber, M. 1969. "Class, Status Groups and Parties." In *Max Weber: Selections in Translation*, trans. E. Matthews, ed. W.G. Runciman. Cambridge: Cambridge University Press.

———. 1982 [1946]. "Characteristics of Bureaucracy." In *Sociological Theory: A Book of Readings*, ed. L.A. Coser and B. Rosenberg, 5th ed., pp. 327–34. London and New York: Macmillan.

Weiss, W.A. 1990. "Challenge to Authority: Bakhtin and Ethnographic Description." *Cultural Anthropology* 5 (4): 414–30.

West, M.D. 2005. *Law in Everyday Japan: Sex, Sumo, Suicide, and Statutes.* Chicago and London: Chicago University Press.

White, M. 1987. "The Virtue of Japanese Mothers: Cultural Definitions of Women's Lives." *Daedalus* 116 (3): 149–63.

———. 1988. *The Japanese Overseas: Can They Go Home Again?* Princeton: Princeton University Press.

Whyte, W.F. 1956. *The Organization Man.* New York: Simon and Schuster.

Williams, R. 1988. *Keywords: A Vocabulary of Culture and Society.* London: Fontana Press.

Wittgenstein, L. 1997 [1958]. *Philosophical Investigations*, trans. G.E.M. Anscombe. Oxford, UK, Cambridge, MA: Blackwell.

Wolf, M. 1972. *Women and the Family in Rural Taiwan.* Stanford: Stanford University Press.

Wolf, N. 1990. *The Beauty Myth: How Images of Beauty Are Used Against Women.* London: Vintage.

Wright, S. 1994. "Culture in Anthropology and Organizational Studies." In *Anthropology of Organizations*, ed. S. Wright, pp. 1–31. London and New York: Routledge.

Yamashita, E. 1996. "戦後社会と女性－職場と家族の変容" (Postwar Society and Women: Changes to the Workplace and Family). In 溶解する女と男 21世紀の時代に向けて―現代 (The Dissolution of Woman and Man Towards the 21st Century—Present Day), Vol. 7, 女と男の時空―日本女性史再考 (The Space-Time of Women and Men—Reconsidering Japanese Womens History), ed. E. Yamashita, pp. 599–641. Tokyo: Fujiwara Shoten.

Yoda, T. 2006. "A Roadmap to Millennial Japan." In *Japan After Japan: Social and Cultural Life from the Recessionary 1990s to the Present*, ed. T. Yoda and H. Harootunian, pp. 16–53. Durham and London: Duke University Press.

Yoshino, K. 1989. *Cultural Nationalism in Contemporary Japan: A Sociological Enquiry.* London and New York: Routledge.

Zielenziger, M. 2006. *Shutting Out the Sun: How Japan Created Its Own Lost Generation.* New York: Nan A. Talese.

Žižek, S. 2005 [1994]. *The Metastases of Enjoyment: Six Essays on Women and Causality.* London and New York: Verso.

———. 2006. *How to Read Lacan.* London: Granta Books.

Newspaper Sources

Asahi Shinbun. 22/10/06. "増えるシングルマザー：30~40代前半で76万人 (Single Mothers on the Increase: 760,000 among the thirties to early forties)." あっと！@デ゙ータ (Wow! @Data, Life+Science).

Chūnichi Keizai Shinbun (CKS). 11/08/98. "人事・労務全般を一括管理：トヨタが新システム" (Managing Overall Information Concerning Personnel and Labor Issues in One System: Toyota Implements New System).

Guardian. 27/03/2000. "Can Snail Mail Beat Email? Electronic Messaging Hasn't Revolutionised the Office Yet."

———. 07/06/2000. "Spreading the Messages: Soon They Will Take Up the Whole Office Day."

Nihon Keizai Shinbun (*Nikkei*). 14/06/97. "均等法改正 岡野労相に聞く：時間外労働指針で歯止め" (Interview with Okano Labor Minister about EEOL Reforms: Breaks Applied at Guideline for Overtime Work).

———. 19/07/97. "労働裁判で女性に追い風―賃金・昇進で注目判決" (Favorable Wind for Women at Labor Tribunals—Noteworthy Ruling on Wage and Promotion).

———. 05/01/98. "実力勝負の海外へ、「雌飛」の時)" (Going Overseas Where Competence Determines Victory or Loss, the Time "Females Take Flight") 女達の静かな革命 (4) (Women's Quiet Revolution [4]).

———. 07/01/98. "それぞれの10年、「忠誠」より「やりがい」" (A Retrospective of Ten Years of One's Own, "Self-Will" than "Loyalty"). 女達の静かな革命 (6) (Women's Quiet Revolution [6]).

———. 30/01/98. "百貨店では例年より早くリクルートフェアがスタート：上着は紺色が主流、バッグは大きめ" (Department Stores Start Recruitment Fairs Earlier than the Norm: Navy Colored Jackets Are Mainstream, Bags Tending on the Large Side).

———. 02/02/98. "若い世代で消えた連帯感、趣味や家庭など、社外生活を重視" (Feeling of Solidarity that Disappeared with the Younger Generation, Hobbies, Family, and the Like, Life Beyond the Company Prioritized).

———. 09/02/98. "女性社員比率調査" (Survey of the Ratio of Female Employees). 女達の静かな革命 (Women's Quiet Revolution).

———. 07/07/98. "生産調整、雇用減に直結、「柔軟な労働市場不可欠：労働白書" (Adjustment to Production Levels Directly Connected to Reduced Employment Levels, "An Elastic Labor Market is Indispensable": Labor White Paper).

———. 01/08/98. "人員削減・一時帰休広がる：日軽金－340人減、新日鉄－月2日休業" (Employee Entrenchment and Temporary Layoffs Spread: Nikkeikin-Loss of 340 Workers; Shinnitetsu—Suspension of Work for 2 Days per Month).

———. 06/08/98. "学生のインターンシップ：VBに拡大起業精神学ぶ" (Internships for Students: Extending to Venture Businesses, Learning Lessons in the Spirit of Starting a Business).

———. 07/08/98a. "男性の離職率１２．３％に： 昨年、７４年以来の高水準" (Men's Unemployment Rate 12.3%: Last Year Comparable to High Levels in 1974).

———. 07/08/98b. "大卒就職、最低の６５％： 文部省の今春調査－企業の採用不況で減少" (Employment of University Graduates, Lowest at 65%: This Years' Spring Survey by the Ministry of Education—Reduction Due to Companies Cutting Recruitment).

———. 15/08/98. "インターンシップ 制度、関西企業 相次ぎ導入： 松下・近鉄・関電-即戦力養成の期待も" (Internship Scheme, Successive Introductions by Companies in the Kansai Region: Matsushita; Kintetsu; Kanden—Hopes of Training Impromptu Capacity/Strength).

———. 19/08/98. "人員削減での摩擦回避" (Avoiding Friction Resulting from Employment Cuts).

———. 21/09/98. "女性社員の制服―改正均等法で存廃騒動： 企業、廃止でコスト減狙う 労組、既得権侵害に反発" (Female Employee Uniforms—Revised EEOL Creates Agitation for Continuance or Abolition: Companies Aim to Cut Costs by Abolition; Labor Unions' Backlash against Infringement of Vested Rights).

———. 11/02/99. "３５年目のOL： 多極分化、処遇に戸惑う" (35 Years of OLs: Multiple Divergence, Bewildered by Treatment). 女達の静かな革命 (18) (Women's Quiet Revolution [18]).

———. 23/10/2006a. "企業向上の陰に. 疲する現場、社員ら悲鳴" (In the Shadow of Corporate Improvement. Deterioration On-site: Employees Cry Out). Salaryman Story (604).

———.23/10/2006b."変わる就職、働く私. 広がる女性採用、企業の本気見極めたい、活用策、掛け声より中身" (Changing Hiring Practices: Working Self: Changing Recruitment, Female Recruitment Diversifies—Part 1, Wish to Confirm What Firms Really Have to Offer. Substance Rather Than Empty Words of Encouragement).

———. 21/11/2006. "'幼なじみ'幼少年時代のほのかな恋の記憶." (*"Osana-najimi" a Faint Memory of Love in Childhood*).

Nikkan Kōgyō Shinbun (NKKS). 20/08/98a. "９８就職戦線、９５年の氷河期並み、とくに文系女子に厳しさ" (1998 Recruitment Battle, Equivalent to 1995 Ice Age in Recruitment, Particularly Tough for Female Arts and Humanities Graduates).

———. 17/09/98. "経営幹部育成スクール： 丸紅、若手社員まで拡大、時代の変化に早期対応" (Top Management Training School: Entry Extended to Younger Employees at Marubeni, Swift Response to Changing Times).

Nihon Keizai Sangyō Shinbun (NSS). 03/08/98. "ミクニ、国内の生産体制を再編： ２年計画、早期退職制度も導入" (Reorganisation of Domestic Production Structure at Mikuni: Two Year Plans, Early Retirement Policy Also Introduced).

———. 07/08/98a. "女性だけで販売促進策を企画： トヨタ東京カローラ" (Planning of Marketing Promotion Schemes by Women: Toyota Tokyo Corolla).

———. 07/08/98b. "アサヒコーポ―人員削減計画を達成： 販売子会社8社に統合" (Asahi Corporation—Targets Achieved for Plans on Employee Cutbacks: Marketing Subsidiaries Consolidate to 8 Companies).

———. 12/08/98. "日野自動車―間接部門に「一年帰休」： 異例の経営判断、本格的人員整備を回避" (Hino Automobiles—*Kansetsu-bumon* Given Temporary One Year Layoffs: Unprecedented Management Decision, Avoiding Full Scale Employee Adjustments).

———. 18/08/98. "グループで「横の情報化」―サントリー： 業務を分担、事業効果的に" ("Making Information Available at Horizontal Level" Across the Group—Suntory: Allotment of Work Tasks, Toward Operational Efficiency).

Nihon Keizai Sangyō Shinbun (NSS). 19/08/98. "離職中高年の「才能」登録" (The Registration of Retrenched Middle-Age and Senior Citizen's "Ability/Talent").

———. 20/08/98a. "GHPエアコン55%拡販：女性の営業チーム発足" (Sales Expansion of 55% at GHP Air-Conditioning: Inauguration of Women's Business Team).

———. 20/08/98b. "商社、新卒採用を抑圧：来年度総合職下位3社は3割減" (Trading Firms Hold Back on Recruitment of New Graduates: Low Levels of Recruitment of Career Track Workers at the 3 Lowest Recruiting Firms Set to Decrease 30% Next Year).

———. 16/09/98. "失業サラリーマンの起業支援：通産省が融資制度、主婦・学生にも助成制度新設" (Supporting Redundant Salarymen to Start New Businesses: MITI's Loans Financing Policy, Newly Established Subsidy Policy Also Available to Housewives and Students).

Periodical Sources

Economist. 07/10/99. "The World in Your Pocket." The Economist Survey: Telecommunications.

Nikkei Business. 27/01/97. "女の規制緩和、雇用平等は甘くない" (Restrictions on Women Relaxed, Equality in Employment Is Not Easy).

———. 20/12/99. "トップダウンの組織は解体される：　情報社会こそ、より人間的な社会" (Top-Down Organization Will Become Dismantled: Indeed, an Information Society, Is a More Humanistic Society), p. 39.

Rōseijihō. 13/12/96. "均等法および女子保護規定に 関する意識" (Awareness Regarding the EEOL and the Rules Concerning Protection of Women), Vol. 3284: Sec. 6, pp. 31–5.

———. 24/10/97. "顔でみる新入社員意識の変遷　[編集部作]" (Judging the Changing Attitudes of New Recruits by Their Faces [By the Editorial Board]), Vol. 3324: Sec. 1, pp. 42–3.

Pamphlet Sources

Ministry of Labor, Women's Department. 1998. 均等法が変わります！もう一度職場で女性の雇用管理の点検を (The EEOL Will Be Revised! Request for Re-examination of the Management of Women's Employment Procedures in the Workplace). No. 10.

Osaka Prefecture Employment Department. 1998. 知的障害者雇用促進事例集 (Booklet of the Promotion of the Employment of the Mental and Physically Handicapped). March.

Internet Sources

AGEFIPH. 2008. http://www.agefiph.fr/index.php?nav1=common&nav2=uk, accessed 01/11/08.

Americans with Disabilities Act (ADA). 1990. http://www.eeoc.gov/facts/ada18.html, accessed 01/11/08.

————. Amendments Act of 2008. http://www.eeoc.gov/ada/amendments_notice. html, accessed 01/11/08.

Disability Discrimination Act (DDA) 2005. UK. http://www.opsi.gov.uk/acts/ acts2005/ukpga_20050013_en_1, accessed 01/12/08.

Fox, C.M. 1996. "Changing Japanese Employment Patterns and Women's Participation: Anticipating the Implications of Employment Trends." http://www.soc.hawaii.edu/ con/future/j3/fox.html

Japan Institute for Labour Policy and Training. 2006. Labour Situation in Japan and Analysis: General Overview 2006/7. http://www.jil.go.jp/english/laborinfo/library/ documents/Labor2006_2007.pdf

————. 2008. Table 9–17. Employment Measures for the Disabled. Chap. 9 Worklife and Welfare (chap. available in Japanese only). *Databook of International Labour Statistics.* http://www.jil.go.jp/english/estatis/databook/2008/09.htm, accessed 30/10/08.

Japan Organization for Employment of the Elderly and Persons with Disabilities. 2005. www.jeed.or.jp/disability/employer/employer01.html, accessed 27/07/08.

Ministry of Health, Labour and Welfare. 2002. Section 4. Trends in Human Resources Management Systems, *Part I Trends and Features of the Labour Economy in 2002*, www.mhlw.go.jp/english/wp/wp-1/2–2-4.html, accessed 16/07/04.

————. 2004. Department of Health and Welfare for Persons with Disabilities. www. mhlw.go.kp/english/org/policy/p31.html, accessed 16/07/04.

Ministry of Internal Affairs and Communications. 2000. "Table 1. Employed Persons 15 Years of Age and Over by Occupation (Major Groups) and Sex—Japan." *Labour Force Survey.* http://www.stat.go.jp/english/data/kokusei/2000/kihon3/00/01.htm, accessed 28/10/08.

————. 2005. Women's Workforce Participation by Age. Special Feature. *Womenomics.* http://www.fpcj/.jp/e/mres/publication/flf/pdf/00_womenomics.pdf.

————. 2008 . Table 15. Employed Persons by Occupation and Status in Employment. *Labour Force Survey*, August. http://www.e-stat.go.jp/SG1/estat/ListE.do?lid=000001032687, accessed 28/10/08.

OECD. 2005. *Economic Outlook*, Preliminary Edition, Japan, pp. 80–4. N.d.

Reuters. 01/12/2000. "CIA Disciplines Employees for Secret Chat Room." http://www. reuters.com/news=article.jhtml?type=internet&Repository=INTERNET=REP&Re positoryStoryID=%2Fn01/12/00

Sōrifu. 1996. "The Popular View: A Man's Place Is at Work, A Woman's Place is at Work and in the Home." Advances Made by Women during the 50 Years Since the End of the War, *The Present Status of Women and Measures*, 5th Report on the Implementation of the New National Plan of Action toward the Year 2000. http:// www.sorifu.go.jp/danjo/women96

————. 1998a. "晩婚化の進展 (The Progressive Trend for Later Marriages)." 女性の現状と施策：新国内行動に関する報告書 (The Present Status of Women and Measures: A Report concerning New Behavior in Japan). http://www.sorifu.gp.jp/danjyo/ genjo

————. 1998b. "人口及び人口動態 (The Population and Population Movement). 女子差別撤廃条約実施状況第4回報告" (4th Report on the State of Enforcement of the Agreement to Abolish Discrimination against Women). http://www.sorifu.go.jp/ danjyo/sabetsu/1-1.html

Sōrifu. 1999a. "News From the Headquarters for the Promotion of Gender Equality: April 1999 Ministerial Ordinances and Guidelines on the Amended EEOL," *Women in Japan Today*, January. http://www.sorifu.go.jp/danjo/women99/n.23. html

———. 1999b. Yearbook of Labor Statistics, ILO. In *The Present Status of Gender Equality and Measures.*

———. 1999c. Fig. 3, Changes in Four Factors for Calculating GEM (Gender Empowerment Measure: Evaluates Women's Participation in Economic and Political Activities and Decision-Making) by Country. In *Present Status of Gender Equality and Measures.* http://wwww.sorifu.go.jp/danjyo/english/plan2000/1999/p1c1.html, accessed 17/4/00.

———. 2004. Basic Data on Gender Equality in Japan. *Women in Japan Today.* http://www.gender.go.jp/

———. 2006. 5–2 Women's Participation in Various Fields. 5. Basic Data on Gender Equality in Japan. *Women in Japan Today.* http://www.gender.go.jp/english_contents/women2006/index.html

Statistical Research and Training Institute. 2006. *Current Situation of Parasite-Singles in Japan.* Nishi, F. and M. Kan. (22 March) http://www.stat.go.jp.training/2kenkyu/saika.htm, accessed 07/10/2006.

Waldschmidt, A., and K. Lingnau. 2008. Report on the Employment of Disabled People in European Countries: Germany, pp. 1–10. *Academic Network of European Disability Experts (ANED)*—VT/2007/005. DEEmploymentReport.pdf, downloaded 01/11/08.

Encyclopædia and Dictionary Sources

The Blackwell Encyclopedia of Sociology, 2007, ed. G. Ritzer, vol. 8, s.v. "*Seken*", T. Kurihara, pp. 4154–7. Oxford: Blackwell.

The Concise Oxford Thesaurus, [1995] 1997, s.v. "open," p. 549. Oxford: Oxford University Press.

大衆文化事典 (Encyclopaedia of Popular Culture), 1991, ed. H. Ishikawa et al., s.v. "OL", p. 92. Tokyo: Kobundo.

社会学事典 (Encyclopaedia of Sociology), 1988, ed. M. Mita, A. Kurihara, and Y. Tanaka, s.v. "*seken* (世間)," p. 544. Tokyo: Kobundo.

岩波女性学事典 (Iwanami Dictionary of Women's Studies), 2002, ed. Inoue, T., C. Ueno, Y. Ehara, M. Osawa, and M. Kanou, s.v. "Jiritsu-Independence," Amano, M., pp. 252–3. Tokyo: Iwanami Shoten.

Kenkyusha's Japanese English Dictionary, 1974, 4th ed., s.v. "オープン (open)," p. 1311. Tokyo: Kenkyusha.

新社会学辞典 (New Encyclopaedia of Sociology). 1993, ed. K. Morioka, T. Shiohara, and Y. Honma, s.v. "経営情報システム" (Management Information System; MIS), p. 353. Tokyo: Yuhikaku.

Unpublished Reports and Presentations (Pseudonyms)

Arnaud, final internship presentation, Osaka, Japan, 28/08/98.

"Capucine's Internship over Three Months at the Company," Osaka, Japan, n.d. [c.1997], n.p.

"Ben's Final Internship Report," 13/01/97, Osaka, Japan, n.p.

Performances

Antirom, *RGB Show*, OneDotZero Festival, ICA, London, May 1999. Transcribed from RGB Show, Sydney, Australia, September 1999.

Bubble Trouble 2, television performance, BBC2, 09/01/2000.

Index